Søren Brunak • Benny Lautrup

NEURAL NETWORKS

Computers with Intuition

World Scientific
Singapore • New Jersey • London • Hong Kong

NEURAL NETWORKS
Computers with intuition

1. edition, 1. printing, 1990
Layout: Kitte Fennestad
Cover photo: © Kitte Fennestad
Illustrations: Jesper Tom-Petersen
Translation: Andrew Cameron-Mills

Published by

World Scientific Publishing Co. Pte. Ltd.
P.O. Box 128, Farrer Road, Singapore 9128
USA office: 687 Hartwell Street, Teaneck, NJ 07666, USA
UK office: 73 Lynton Mead, Totteridge, London N20 8DH, England

Original Danish edition published by
Munksgaard International Publishers, Copenhagen

Library of Congress Cataloging-in-Publication Data
Brunak, Søren
[Neurale Netværk. English]
Neural Networks: Computers with Intuition/
Søren Brunak and Benny Lautrup
p. cm.
Translation of: Neurale Netværk: Computere med Intuition.
Bibliography: p.
Includes index.
ISBN 9971-50-938-5. – ISBN 9971-50-939-3 (pbk)
1. Neural computers. 2. Neural circuitry. I. Lautrup, Benny. II. Title.
QA76.5.B78513 1990 006.3 – dc20 89-14671

Printed in Singapore

NEURAL NETWORKS

CONTENTS

FOREWORD

Understanding the human brain is the greatest scientific and intellectual challenge ever faced by man. It has always provoked wonderment, how that lump of nerve cells can produce art and science, war and peace, hate and love. Science is no longer in any doubt that the human spirit has a physical origin in the brain, but nobody has yet made the connection between mind and matter.

In this book we try to describe a new field of knowledge which lies in the borderland between the natural and the synthetic. Natural phenomena have the reputation of being honest, strong and not to be circumvented. Synthetics smack of trickery, "artificial" is a pejorative, and all are viewed with the deep mistrust that we reserve for those things we have ourselves invented. This book also examines the relationship between man and machine and what possibilities these uneasy partners have of ever understanding each other.

With the advent of the computer, machines have attained a degree of complexity which makes it possible to study intelligence in a completely new way. We can now simulate the behavior of small and greatly simplified versions of the neural networks with which Nature has equipped advanced, living beings. These networks have a structure radically different from that of the conventional computer and have shown themselves to be much better at carrying out calculations of a cognitive character.

One of the main themes is therefore the relationship between calculation and intelligence, both artificial and natural, and it is a theme which involves such fundamental existential relationships as language and reality, intuition and symbolism, body and soul. The book is not packed with quotations leaving the responsibility for the arguments to others, but presents our views in a way which makes it possible for the reader to disagree.

Verbal communication is a very restricted channel. We have tried to widen it by evoking in the reader some of the images which have been a part of our thinking. We have not, however, included lots of complicated diagrams of either brains or computers, but have put the emphasis on the principal matters behind computation and on that backdrop of concepts which lies behind the physics in the behavior of the computer.

We thank Johanne Brunak, Wibeke Fonnesberg, Grete Henius, John Hertz, Jørn Hounsgaard, Steffen Petersen, Henrik Oxvig and Morten Olesen for criticism and comment which has helped us greatly with the Danish edition which appeared in October 1988. Special thanks go to our Danish editors Tor Nørretranders, Jørgen Steen Nielsen and Joachim Malling for support and encouragement. For the English edition we owe thanks to Johanne Brunak, John Hertz and Choy Kam Luen.

SØREN BRUNAK AND BENNY LAUTRUP
Copenhagen, August 1989

PROFESSION: COMPUTER

Dutchman Wim Klein was a computer.

On August 27, 1976 he extracted in 2 minutes and 43 seconds the 73rd root of a 500-digit number — in his head! In the field of mathematical wizardry, normally regarded as the exclusive realm of the electronic computer, Wim Klein was an impressive magician. Unlike electronics, his performances were not entirely silent, but were accompanied by a muttered mixture of Dutch numerals and profanities.

The term *computer* was originally the name given to someone who performed calculations for a living. Computers typically worked for banks or insurance companies, where loans and insurance policies required extensive calculations. When this work was taken over by machines, they also took over the name.

Calculating is not just juggling with figures. When the human brain constructs, in a tenth of a second, a three-dimensional experience from two-dimensional images imposed on the retinas of the eyes, a very complex calculation is taking place. Such calculations are, in fact, so complex that it would take a supercomputer several hours just to sort out the spatial relationships which occur in traffic at an ordinary intersection. What is more, the human brain's calculations are so certain that it is possible, for example, for many people to drive simultaneously in traffic at speeds far higher than those we are actually designed for.

The brain can process formidable amounts of information. Through the eyes, ears, nose, mouth and skin, information pours into our brains. The so-called "five senses" are not the brain's only input channels. Information about the state of the body and its internal organs is constantly being transmitted to the brain. Various parts of the brain are incessantly sending each other messages: memories, status reports, decisions, alarms and what not.

This colossal amount of information is processed and transformed partly as output observable to the outside world in the form of muscle movements and partly to the brain itself in the form of thoughts. A pianist carries out highly organized muscular activity, largely governed by input from the senses. Conversely, the passive experience of listening to piano music is an inner mental process whereby input from the senses is translated to thoughts and feelings.

Calculation is not just mathematical work in the ordinary sense, but includes all forms of information processing. A calculation is a process that transforms one set of data, the input, into another set of data, the output, which is the result of the calculation. Today, machines which perform this function are called computers.

Contrary to what might generally be supposed, these processes never *create* information. Normal information processing actually works on a principle of forgetting, or more accurately, discarding information. The tricky thing is, of course, to do it in the correct way. Consequently, these processes cannot normally work backwards, just as it is usually impossible to reconstruct a question on the basis of an answer.

In the brain, the end product of a computation is the focusing on a particular piece of information. The famous "cocktail party effect" is an example of the brain's incredible ability to focus and discard information. When a person is standing in the middle of a large group of talking people, his brain can single out a particular conversation from all the others, for example that between his wife and some sexy-looking Jack Nicholson type. All the rest of the incoming information — the other conversations — is discarded and there is no way of reconstructing it from the conversation between the wife and the satanic satyr.

All forms of cognition are based on *reduction* of information. Without the use of such rough concepts of classification as truck, contract, temperature, or politician, one would never be able to make coherent associations and would experience the world as a swirling chaos of isolated details. The brain boils all these details down to an understandable form which can be both comprehended by oneself and communicated to others.

During the '40s and '50s Wim Klein, the computer, performed in French and Dutch circuses. In 1958 he finally got a steady job at the European Nuclear Research Centre, CERN, in Geneva. His function was to perform complex calculations for theoretical physicists, using his phenomenal numerical ability. However, within a few years he was outsmarted by a new generation of computers, the speed of which had vastly increased and which, unlike Wim, could run nonstop 24 hours a day. But, for special types of calculation, he was still supreme and later gained a place in the *Guinness Book of Records* for the incredible ability to extract roots mentioned at the beginning of the chapter.

Today, machines have outdistanced us all in the domain of "number crunching" and related symbol manipulation. Wim Klein and his colleagues have now been transferred to other types of work, and his 1958 "victor's crown" has been taken over by the electronic computer and will never again be won by human brains. The digital computer is, however, hopelessly inferior to humans in almost all other respects.

Human intuition rapidly solves complex problems which conventional computers and their programmers are quite powerless to deal with. To recognize someone from a mere glimpse of his face involves data processing of such speed and extent as to make the fastest computer seem like a snail. When we recognize a face, it occurs almost momentarily, and it is remarkable that we are completely unable to consciously halt the process.

The brain and the run-of-the-mill computer have widely different architectures, and this markedly affects the type of functions each is best able to perform. No matter which computer you buy — from a miniature pocket calculator to a giant supercomputer — they all process information more or less after the same principle: one thing at a time. Although most computers are so fast it seems that many users are served simultaneously, this is actually an illusion. The computer in fact allocates each user a tiny time slice and moves constantly around between them.

In contrast, the human brain can conduct an enormous number of operations simultaneously and in parallel. Information processing is carried out by a network of connected nerve cells called neurons. This network is composed of an astronomical number of neurons and is vastly more complex in structure than the telephone network of the entire planet.

The work of the conventional computer is fully centralized, whereas that of the brain to a much greater extent is done through decentralized, distributed influence instead of control. The two architectures are as ideologically different as the structures of totalitarian and anarchic states.

These ideological differences in structure are clearly reflected in their different vulnerabilities. Everybody knows just how fragile and sensitive computers are. At the least provocation they go "down", whereas the brain — even under the most wretched conditions — is "up" year after year, without interruption through an entire lifetime.

The dream of artificial intelligence has for many years been based on the centralized structure, but after 40 years of research this dream has not come true. It seems as if imitation of the natural intelligence of living organisms will not succeed without borrowing ideas from the structure of biological neural networks.

For centuries, the heart was supposed to hold the secret of man's divine origin, be the seat of his soul. Modern medical science has destroyed this myth and placed the heart in perspective as being an organ similar to the pump in an ordinary washing machine. The Norwegians, for example, have taken this insight to its logical conclusion in their language and simply call the organ a "bloodpump"!

Having been driven out of the heart, the soul took up residence in the brain. This diminution of the heart's "divine" status has not, however, made it any easier to be unhappily in love. Heartache is still synonymous with the pain of love, and even if the heart is now no more than a pump, the human emotions have not yet been reduced to the mechanical performance of a washing machine.

The traditional attitude of medical science towards the mind and the brain has been — and still is — marked by the father of modern anatomy, Niels Stensen, in Latin Nicolaus Steno. He died in 1686 and was canonized in 1988 by the Catholic Church because of his ability to operate remotely — after death — as a miraculous healer. Steno held that there were fundamentally two approaches towards understanding the brain. Either one could ask the divine Creator, or one could try to dismantle the brain into its basic parts to see how it worked. He felt that the first approach was rather impractical, as direct communication had been discontinued for some time ("God is alive, but He doesn't want to get involved"), and, therefore, the second approach was the only realistic possibility.

Brain research has not had much success in identifying the exact location of the mind. Regardless of how much the anatomy and physiology of the brain have been studied, nobody has located the mind yet, even though most agree that it must be in there somewhere. The reason for this could, of course, be that the mind is not to be found in any particular place, but is distributed over the entire network of neural calculators that makes up the brain.

It is a common belief that if one understands the functions of the parts in sufficient detail, it is an easy matter to deduce the function of the whole. The fact is, however, that in nature, *collective effects* of many interacting parts often show up, effects that cannot be explained easily in terms of the properties of the individual parts.

We are alive not because each individual biomolecule in us is alive. Life is not the sum of small lives, but an irreducible property of large molecules interacting collectively with each other and their surroundings, a property which cannot be broken down into corresponding properties of its component molecules. Otherwise detergent with its component of biological enzymes would also be alive.

What we know as consciousness, mind, and intelligence must also be collective effects of the interaction between many neurons, effects that cannot be reduced to similar properties of individual neurons. The construction of artificial neural networks is a modest attempt at understanding collective effects occurring in such networks. It is still an open question whether one will be able to manage with much smaller networks than those provided by Nature. Artificial neural networks are media for knowledge, and the study of them attempts to decipher the enigmatic means by which Nature carries out information processing.

WHOLESALE RIDDLES

Man is a curious creature, possibly the only animal that attempts to acquire a deep understanding of his surroundings. The need to understand is not dictated just by a desire to control one's circumstances, but can be purely esthetic or even religiously motivated. We have a compulsion to understand what the universe is, what we are, and finally why we ask why. We live in a mysterious world full of great riddles.

The greatest riddles concern the structure of the universe, the nature of matter, the essence of life, and the necessity for the mind. These are all "wholesale riddles", each involving millions of detailed questions of which few have yet received satisfactory answers.

Throughout history, all these subjects have been taboo at one time or another.

In 1633 Galileo was threatened with execution if he would not recant his support for the theory that the Earth moved round the Sun. The prevailing opinion at that time was that the Earth was the center of the universe which was composed of a number of concentric spheres with God residing in the outermost. The Catholic Church had even set up a "subsidiary company", the Inquisition, to safeguard the purity of the faith. This organization regarded the protection of the geocentric view as part of its objective.

Alchemists were the first experimental chemists. Their search for gold ensured them a certain position with both secular authorities and the Church, both of whom were regularly in pressing need of that particular substance. However, alchemists were always in danger of being accused of witchcraft and having more than just their fingers burned!

Darwin's ideas about the origins of life on Earth were received with great scepticism and opposition. When Queen Victoria was told that we were descended from apes, she said that she hoped it was not true, but if it should be, she hoped that it would not be commonly known. Even in the United States today, there are many places where parents insist that the schools teach their offspring the biblical version of creation and evolution in biology classes.

The soul and the origins of thought have frequently been taboo subjects: souls belong to the Church and thoughts are part of the manifestations of the soul. The workings of the brain are still much more of a mystery to us than are other biological phenomena — thought processes have a radically different character than bodily metabolism. If one is tempted to use a scientific approach to the riddle of the necessity for the mind, one can still be accused of heresy and ridiculed.

THE STRUCTURE OF THE UNIVERSE

If one is interested in the question of why the mind is necessary to man, one cannot avoid examining the relationship of man to the universe.

In the Western societies there are not many people left who still believe in the biblical version of creation, even though in most respects it is very easy to digest: An extremely efficient elderly gentleman set the whole thing up and got it going in the space of one week. The work was carried out in the grand old preindustrial tradition: by the master craftsman singlehandedly! After which he mostly kept his nose out of things until his son could take over the business.

Modern science, however, works with a theory of creation which, in contrast to the cocksure certainties of the Bible, has no definite answers to most of the deepest questions.

The *Standard Model* of cosmology postulates that the universe began about 15 billion years ago with a "Big Bang". Fragments from this gigantic explosion are still moving away from each other. They later became galaxies, stars, planets and us. Today we exist in a very cold and expanded universe which was originally very hot and compressed. The universe will either continue to expand or, at some point, the process will reverse and it will contract again, culminating in a gigantic implosion. We have as little idea about what was before the explosion as what will come after the implosion. In such extreme situations, the concept of time, as we use it, may have little or no meaning.

According to the Bible, mankind occupies a central position in the universe, entirely in keeping with our characteristic self-admiration. In contrast, three hundred years of cosmology have given us the understanding that the universe would continue to operate perfectly without our presence. It is humiliating to have to acknowledge that man's position in the universe is to live out an insignificant existence on an insignificant planet, near an insignificant star in an insignificant galaxy.

THE NATURE OF MATTER

The matter of which we are all composed has a microscopic structure reminiscent of a series of Chinese boxes. Each time the "magnifying glass" is made stronger, a new, smaller structure is revealed inside the old. The deeper we penetrate into the nature of matter, the more complex and alienating its structures become. From the atom downwards, they can only be described with the use of advanced mathematics. We have no direct experience of Nature at these levels and are therefore denied the usual use of our intuition and must resort to the formalized methods of mathematics.

Matter, like all good technology, is based on modules or building blocks. All chemical processes use building blocks called molecules. Molecules are, in turn, composed of other modules, atoms, which are themselves composed of elementary particles. In spite of their name, elementary particles are anything but elementary, and are composed of even smaller modules. At the deepest level, the modules are perhaps the latest hit in the study of matter, "superstrings".

If one naively inquires what superstrings are made of, one is tumbling into a realm of comprehension similar to that involved in the question of what existed before the universe began. If there is indeed some level of structure below superstrings then the question simply spirals down in a dizzying way. Religions resolve all these questions with the irrefutable concept of *faith*, successfully designed to stifle curiosity.

Perhaps such "tail-biting" barriers to comprehension merely reflect the way in which our brains strive to make sense out of sensory impressions. Farthest out in the universe or deepest in the heart of matter we may only see reflections of our own minds!

THE ESSENCE OF LIFE

What is in fact living and what is dead is no trivial question. In the Period of Enlightenment one could in disputes "prove through pure logic" that a stone was alive, since a stone cannot fly and since humans are alive and cannot fly, but such arguments do not cut much ice with us today.

It is clear to us that we are alive as, for example, are dogs. Looking down into a microscope, one is quickly convinced that bacteria are also alive, and we can surely also agree that stones and atoms are dead.

A virus, which causes a cold, is an organism that makes use of our private cell machinery to replicate itself. Viruses do not eat us; they are party crashers that merely sponge on our equipment. Some viruses are so stupid that they kill us, thus destroying their own means of existence. It is rather a matter of taste, whether viruses are alive or not.

Life is much easier to define in terms of processes than of properties. One of the most central processes of living organisms is the ability to reproduce. Today, we can make machines which copy certain objects, but they cannot copy themselves. Nobody has yet seen a big xerox machine spit out a little xerox machine in the tray. In order to make xerox machines you must get hold of a factory, a kind of "meta xerox machine" plus the recipe for how to make them. And xerox machine factories cannot themselves produce xerox machine factories. New factories are designed and built by workers, functionaries and shareholders. But who makes these people? Their parents, who were themselves workers, functionaries and shareholders. In this sense the xerox machine is an evolutionary dead end. It does not critically participate in its own reproduction at any level.

Cells are alive because they master the fundamental processes of metabolism, growth and reproduction. Cells were originally separate entities which took nourishment from the inorganic lukewarm chemical soup that they cruised around in. It took a couple of billion years for these single cells to associate into multicellular organisms.

Cells contain two principal components, information and machinery, the interaction of which makes precise self-replication possible. The greater part of the information is stored in long chains of nucleotides, called nucleic acids or DNA molecules. In these molecules the basic genetic messages are formulated in a language written with four letters, one for each of four different nucleotides. The machinery is mainly composed of proteins which carry out metabolism, including such specialized cell functions as communication of neural signals. Part of the machinery is capable of building new proteins on the basis of the information contained in the DNA chains, and these proteins are incorporated in new machinery. DNA is the book of recipes from which all proteins are prepared. There is also cell machinery capable of reproducing DNA and thereby the messages it contains. Life, as we so far understand it, is a complex interaction between information in DNA chains and the protein-producing machinery of cells.

Life is not what it used to be. It presumably started on Earth nearly four billion years ago, not long after the planet itself was formed. The mystery behind the beginning of life on Earth is not solved. But once a self-replicating process starts, it can be very difficult to stop it. If several processes are simultaneously active, they will compete for the available reactive components. Competition in speed of reproduction will lead to Darwinian selection, with the result that the processes are refined and attuned to each other. If some of the processes work together and thereby increase their joint speed of reproduction, more and more self-reinforcing patterns of chemical reaction can arise.

In his book *Origins of Life* Freeman Dyson has proposed that the two cellular components in fact stem from two independent life forms in symbiosis. This theory would mean that at the birth of life on Earth there was only the cellular protein machinery without the information-carrying DNA. This cell machinery could exchange matter with the environment and divide so that the number of cells steadily increased. At some time these basic cells incorporated a new life form, DNA-related RNA, which, like viruses, could utilize the cell machinery to reproduce itself far faster within the cell than outside. In time, the cell machinery "realized" that it could utilize this method of replication to make its own process of cell division more uniform and precise. From being a "disease", which utilized the facilities of the cell like a virus, RNA became a "memory", a record of the cell's "good ideas". Many times in human history the assimilation of foreigners into an existing native population has been to mutual advantage. It could well be that a corresponding process occurred in the earliest history of life on Earth.

Life itself has also affected the stage on which life's history is enacted. For example, Earth's atmosphere originally contained no oxygen, thus oxygen-utilizing creatures could only occur after plants had created an appropriate environment by the production of oxygen. And mankind is now making "holes in the sky" by means of deodorant sprays!

It is not possible to work back in a direct line from the cell chemistry of current living organisms to that of the original chemical soup. Far too many blackboards with important intermediate results have been erased. Still, when we study the essence of life, we do not face a philosophical barrier as we do in studying the structure of the universe and the nature of matter. We know that atoms are not alive, but that cells are. Somewhere along the line, between atoms and cells, the transition between what is dead and what is alive occurs.

The evolution of computers which we are now witnessing also wipes the slate clean of intermediate forms. It is hard today to infer from a credit-card-sized pocket calculator jammed with transistors that once computers were made with vacuum tubes and were monstrously large. Evolution, be it natural or artificial, often employs a "ladder" to crawl up to the next level, and then throws away the ladder with a disdainful shrug. The evolutionary extravaganza of invention is so rich that Nature — and we — can afford to throw away many ladders after only short use.

In 1987, at the Los Alamos National Laboratory in the United States, prize money was offered to the first person to start an autocatalytic, self-replicating reaction which would continue to generate complexity, resembling artificial evolution. Nobody has yet determined the necessary and sufficient conditions for such a self-organizing process. However, even though the essence of life is difficult to penetrate, the very fact that such a prize was offered indicates that even though the problem represents a great riddle, it is in fact a soluble one.

THE NECESSITY FOR THE MIND

In some way or other it would be very strange if the universe were simply there without anyone discovering that it existed. It may have been so for a very long time. We *could* be the first to reflect on existence.

This fundamental existential question has vexed mankind since the beginning of time and driven him to invent religious systems with beings at higher levels than the earthly. Gods are a type of guards, keeping watch even when we are not around. But who guards the guards? This is perhaps an even more troublesome, vexatious question than the first.

Instead of trying to answer these existential questions, let us, for the sake of argument, accept that we have a mind and that evolution has managed to create it. That is more than just evolution — that is revolution. Nature must have had a great empty niche for minds, otherwise this revolution would not have occurred. It is not easy to identify what this niche was, but the mind does not, cannot, exist only for the sake of our bright eyes.

In Nature's ecosystems, species compete for existing niches while simultaneously establishing new ones. As ecosystems grow in complexity, they tend to include more redundant information and are therefore more robust against the various influences to which they may be exposed. Evolutionary niches have parallels in the economic niches of society, where new products and discoveries can dramatically change existing conditions.

Verbal communication is a well-developed human ability and a prerequisite for the organized society in which humans coexist. Possibly, that niche in which the mind emerged was characterized by selection of individuals on the basis of their capacity to communicate. Perhaps the mind is just a spinoff of a development which aimed at an effective means of communication.

Communication takes place with the help of symbols,

in the human context, principally by means of sounds and, in the last few millennia, written characters. This presupposes the existence of some sophisticated symbol-processing apparatus in the brain, an apparatus apparently lacking in those animals which preceded man. The developmental pressures must have been enormous. Within a couple of million years man has achieved a fantastic ability to process symbols in addition to his existing ability to intuitively process information in nonsymbolic form. The major part of our brain is still used for nonsymbolic processing, such as the handling of sensory impressions and motor control, but on top of that we have developed a domed forehead to take care of those higher cognitive functions which symbol processing relies on.

Without these two forms of information processing we would not be humans. Similar to that symbiosis existing between the information-bearing DNA and the protein-making cell machinery, there is a symbiotic relationship between these two totally different information processing aspects of the brain. Perhaps in the same way as life, as we define it, is a product of symbiosis between machinery and information in our cells, so the mind is a result of symbiosis between symbolism and intuition.

Possibly the ability of the mind to reflect upon itself is similar to the ability of living organisms to replicate themselves. In both cases it is a cyclic process. These cyclic processes appear to come into effect only when the complexity of the systems in which they take place exceeds a certain level. Living organisms cannot be composed of just a few atoms and a mind cannot arise in a system with only a few calculating units. A small worm with only 336 nerve cells has not any significant amount of mind.

Perhaps the analogy between life's self-replication and the mind's self-recognition is not just coincidental, but is related to the fundamental characteristics of truly complex systems.

Man may also be characterized as the first animal that has turned the relationship between body and brain upside down. For all other animals the body is dominant, the brain is a service organ for the body, ensuring that its needs are fulfilled in a coordinated way. In man the opposite is the case: the body is the service organ of the brain, ensuring that it reaches the places it wants to reach and plays the games it wants to play. Suicide, the brain's destruction of the body and, thus, of itself, is virtually unknown in the Animal Kingdom. No matter how badly a dog may be treated, it does not on purpose jump out of the nearest window. Animals in fact do not seem to know about death at all.

The tale of the mass suicide of lemmings is a hoax. From time to time lemmings undergo a population explosion, and attempt to migrate to new territory. During these migrations, they cross smaller stretches of water, but when they reach the Atlantic and try to swim to America they drown in the attempt. It has been claimed that the lemmings once photographed by Walt Disney, had to be hounded into the water by the cameramen.

In most countries, brain death is the new criterion for human death. Most people believe that the soul leaves the body and the mind disappears when the brain dies. A body without a mind is a drooling vegetable which can at best only serve as an organ bank. That today, in contrast to such earlier peoples as the ancient Egyptians, we are to a large extent personally indifferent to the eventual fate of our bodies may be judged from the fact that most people are happy to bequeath their organs to post-mortem transplantation.

Man's self-appointed divine status stems from the brain's dominance of the body. It is therefore judged blasphemous to attempt to understand the function of the brain. While modern biology eventually managed to demystify the essence of life, the advocates of the magic of the mind's necessity remain alive and well.

Few people have anything against discussion of such basic physiological brain components as nerve cells, cerebrospinal fluid, neurotransmitters, dendrites, axons, synapses, ion pumps, and other bodily goods, particularly if it is in the context of curing illness. Likewise, few are averse to discussion of the psychological aspects of the brain's function, whether in the context of illnesses of the mind or as a topic for evening school classes.

But should one venture into the realms of the relationship between the neurological and the psychological, one is on forbidden ground. Outside professional circles it is forbidden to examine the causal relationship between matter and spirit, body and mind, neurology and psychology.

That a causal link must exist between matter and mind is more and more part and parcel of the scientific attitude. Every thought must have a physical correspondence with neural activity in the brain. A divine explanation of the connection between matter and mind is not acceptable because it seals off our curiosity. The relationship between mind and matter is a considerable mystery, but it is not pure magic.

THE BRAIN'S WETWARE

Wetware is a technical term used to describe the soft tissue components of the brain, its biological and electrochemical circuitry. This incredibly complicated system is the result of fierce evolutionary competition. A billion years ago living organisms had no nervous systems. Today, we all waltz around with a brain which is more complex than anything else in the universe — at least in the universe we know of.

Not only does the central nervous system function in its present form, but it has also been functional throughout evolution. In contrast to a car, for example, which won't function until the last washer is in place, the nervous system has evolved gradually and with ever-increasing sophistication. This is the way all living organisms evolve. Improvement can only take place on top of an existing system. The complexity of living beings is simply too high for Nature to start from scratch again and again. On the other hand, if a property is lost, it is gone forever. An extinct life form never returns.

This evolution by small steps, which is the result of the need for continuing stability, means that all living organisms include many elements of earlier projects. The cerebellum, which is believed to participate in co-ordination of sensory input and bodily movement, is an example of just such a project which still exists, more or less, in the same form it received millions of years ago. Thus, even though the brain in its present form is incredibly effective, it is surely not the only way it could be organized, if one were to start from scratch.

It is therefore not surprising that such a system is very difficult to find one's way around in. Over the last two centuries the brain has been mapped in ever-greater detail, and today the collected body of information is overwhelming — and almost equally difficult to get around in. All the sensory organs and the motor apparatus have been studied in depth. The peripheral nerves, their paths from the sense organs to the brain and from the brain to the muscles have also been mapped. Their pathways can be traced quite a long way into the brain, and we know where they join other nerves and where they end on the surface of the brain, the cerebral cortex.

However, what we still lack is a fundamental understanding of the interplay between the senses, mental integration and motor control, between experience, thought and action. We know which centers of the brain are involved, but have so far not managed to work out much about the actual nature of the interactions. We know very little, in fact, about how information is processed in the brain, but one thing is clear: information is processed by a network of nerve cells, a *neural network*.

In the neural network vastly different types of perception are translated into a common form, but what that is, nobody knows precisely. The network blends smell, hearing and sight into a neural "cocktail" which provides us with coherent information about widely different physical phenomena. There is significant difference between the chemical reactions which occur in the nasal mucosa, the mechanical movements in the inner ear, and the light-sensitive processes in the retina. But they all end up as patterns of neural activity.

To make a real cocktail, a bartender mixes such ingredients as tomato juice, vodka, a little salt, pepper and tabasco. After they have been well stirred or shaken these ingredients now constitute a new entity called *Bloody Mary* which no bartender could separate out again into its component parts.

Before we experience the cocktail as the complete entity it is, the brain has separated it out in another way. The individual neural activity patterns from each of the senses — smell, taste, sight, and the clinking sound of ice — have merged together to finally result in an experience that produces a sigh of satisfaction.

Is there some supervising bartender in the brain that mixes all the signals again? Who sighs? Is there in fact anyone home at all?

Probably not. Nobody has ever managed to locate a particular spot in the brain which is the seat of the mind. It looks very much as if it is evenly distributed over the whole neuronal net with no one part dominating. It is very difficult to imagine such an absence of localization. Many of the things which have occupied human speculation up to now: for example genes, molecules and atoms, can be dismantled and the meaning understood from the parts, but the mind is as ephemeral as the mists of a summer night.

MEASURES

The human brain has a volume of approximately 1.5 liters, but this size varies considerably from person to person. It is principally made up of water and therefore weighs about 1.5 kilograms. The brain consists of white matter packed in a gray wrapping, and vaguely resembles crumpled newspaper. However, unlike normal packings, as far as the brain is concerned, the wrapping is of the highest importance. It is its surface, the cerebral cortex, which is responsible for the brain's phenomenal information-processing capabilities.

The cerebral cortex has an area of about 2000 square centimeters, the same as that of an ordinary chessboard. A sphere with a volume of about 1.5 liters has a surface area of only 634 square centimeters. Thus, it is the convolutions of the cerebral cortex with its characteristic shelled walnut appearance, that give it a surface area three times greater than it would have if it were smooth.

Cortex is about 3 millimeters thick and is almost everywhere composed of six densely packed layers of neurons. As the neuronal cell bodies are only a few microns in diameter, it means that the cortex must contain some 100 billion neurons, which is about the same as the number of stars in the Milky Way! The central nervous system as a whole has perhaps 1000 billion neurons, which can be compared with the number of grains of sand in a cubic meter of beach.

The neurons are connected by long, thread-like links which are, in reality, extrusions of the cell bodies. The white matter of the brain is made up almost exclusively of these connective threads and their insulation material. The white color of this insulation material is due to a fatty substance found in special support cells. There are ten times more support cells in the brain than there are neurons. Even though in most cases they act as humble servants, their job also includes starting to eat the neurons immediately if any of them should die. Cell death is a very ordinary occurrence. In a child's first years more than 15% of its nerve cells die because they never manage to establish useful connections. There is no social welfare system in the brain.

Even though the neurons are linked together they are not connected like Siamese twins. Each neuron has its own identity, but meets others at certain contact points called *synapses.* At these synapses, signals are passed from one neuron to another, and each neuron is connected to between 1000 and 10 000 others. The number of connections varies greatly both from place to place in the brain and between different cell types. In the cerebellar cortex there is a type of nerve cell which communicates with up to 200 000 others. All in all, the brain contains something in the region of one million billion (10^{15}) synapses.

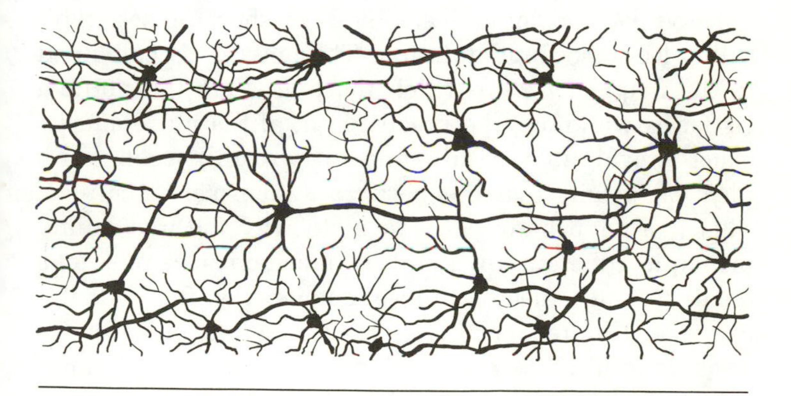

Figure 1: A piece of neural network from the cerebral cortex. Only about 5% of the neurons present in this fragment of the network can be seen in the picture. The rest have not reacted to the contrast stain used to enhance their appearance.

The brain has an energy consumption of about 20 watts, about 20% of the body's total. Energy consumption is remarkably constant, though it falls during sleep, except for dream periods. During dream sleep, the brain can use more energy than when it is awake, which indicates that dreaming serves an important physiological purpose. One theory of dreams, which connects it to the function of the neural network, will be described in the chapter on binary neural networks.

The surface of the globe is spanned by a gigantic telephone network which enables almost every phone owner to ring almost everyone else. Today, there are about one billion telephones, but they are much more sparsely connected to each other than the neurons in the brain. Whichever way you look at it, the brain's network is frighteningly complex compared with the already staggeringly complicated ramifications of the world's telephone network.

Nerve cells communicate by very short pulses, typically of milliseconds duration. The message is contained in the frequency with which the pulses are transmitted. As far as we know, the precise form of the pulse and its exact arrival time within one to two milliseconds are physical details that the nervous system cannot utilize. This is a contributory reason as to why the uncertainty principle inherent in quantum mechanics cannot take credit for such macroscopic phenomena as free will, for example. The frequency can vary from a few pulses up to several hundred per second which is a million times slower than fast electronics, but as there are 10^{12} neurons in the central nervous system, the maximum communication speed may be more than 10^{14} pulses per second. This is the brain's *bandwidth* which is certainly never exploited to its full extent. In fact, only a very small fraction of the neurons are ever active at any given moment.

One can try to assess how many such simple arithmetical operations as addition and multiplication the biochemical processes of the brain could accomplish in one second. Such an assessment is only a handwaving argument and may easily be a factor of 100 too small or too large. Nevertheless, such estimates can provide an impression of the brain's truly formidable capacity.

If we assume that it takes ten arithmetical operations to handle the input that one neuron receives from another via a synapse, we are not being extravagant. Anyway, it is difficult to suppose it could be handled by less. Then if every neuron is receiving input from 1000 others, this would mean that, together, the whole thing would require 10 000 operations. But as this signal integration can occur approximately 100 times per second, each neuron can perform calculations corresponding to about 1 million arithmetical operations a second. As there are 10^{12} neurons in the central nervous system the total equivalent capacity would equate to 10^{18} operations per second. This is more than *100 million times faster* than any of today's computers.

The brain is constructed during gestation and has by and large all its neurons at birth. However, many of the interneuronal connections have not yet been made and are only developed through the influence of experience. The human embryo stage lasts — as most people are aware — for 9 months, and were the brain to develop at an even rate it would mean that somewhat more than 4000 cortical nerve cells were created every second. Naturally, the process is not linear but more nearly exponential, such that it goes faster and faster. The number of

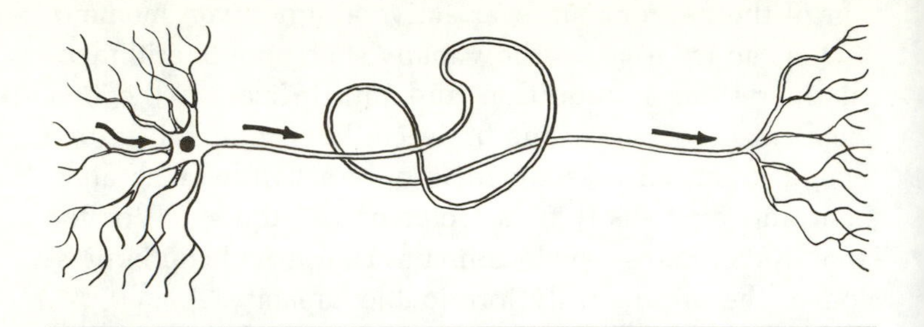

Figure 2: A biological neuron. A neuron is composed of a cell body and two types of branches: axons (transmitters) and dendrites (receivers). The cell membrane contains pumps of various types that can maintain imbalances in charge concentrations inside and outside the cell, which make possible the transmission of electrochemical impulses through the neural network. In the inactive state the inside of the cell possesses a constant negative charge level. A change towards positive charge level can be forwarded — like an impulse — by successive changes in the permeability of the cell membrane.

synapses in the fetus is perhaps 100 times greater than the number of neurons, and the creation rate could average 400 000 per second.

BIOLOGICAL NEURONS

So the nervous system, just like all other bodily organs, is composed of cells. Like other cells, nerve cells have a nucleus which contains hereditary characteristics and a plasma containing molecular equipment for the production of material needed by the cell. As can be seen in Figure 2 the form of nerve cells is highly surrealistic,

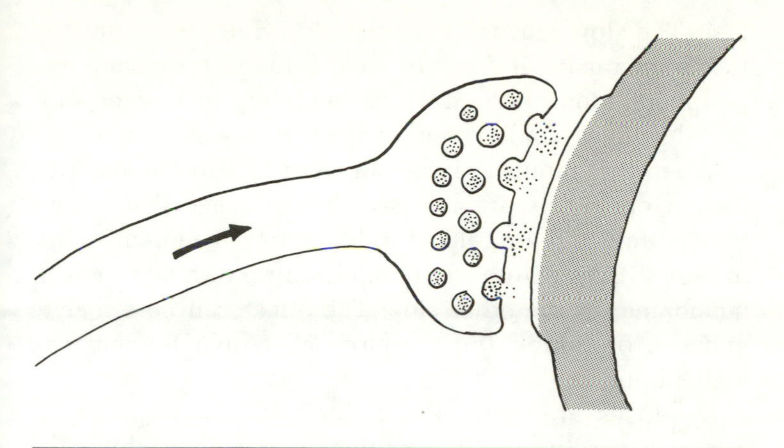

Figure 3: Diagram of a synapse. A synapse is a point of contact between two neurons. The synaptic junction or gap measures about 0.1 micrometers. The neurotransmitter is contained in small spherical vesicles and is released into the junction by fusion of the vesicles with the membrane. It takes something of the order of 0.1 millisecond for the neurotransmitter to diffuse across the synaptic junction and activate the receptor in the receiving cell. Depending on the type of neurotransmitter it will make the receptor cell membrane either more electrically positive (excitation) or negative (inhibition).

distorted in the extreme, with tentacles outspread like the branches of a tree, to make contact with many other cells. There are basically two types of these tentacles, one sending, the other receiving signals — one-way communication channels known respectively as *axons* and *dendrites*.

Neuronal impulses are transmitted outwards from the cell body along the axons to terminals in the synapses. The impulses are electrochemical changes which are transmitted without significant weakening at a rate of up to 100 meters per second. Transmission requires energy and is reminiscent of the glow of a lighted firecracker fuse. The glow moves so fast that the whole fuse lights up instantaneously. It is only in the longest sensory and peripheral motor nerves, which can be up to a meter long, that the effect would resemble the running glow of a fuse.

When the impulse reaches the terminal in the synapse, certain chemicals are released. These are called neurotransmitters and are related to hormones. The neurotransmitters diffuse across the synaptic gap and affect the cell membrane on the other side. The effect can be either to enhance or inhibit the receptor cell's own tendency to emit impulses.

Synapses are normally divided into two principal groups according to whether they enhance or inhibit the receptor cell's impulse emission rate. Synapses can, however, be very complicated and there are many examples of hybrid forms which combine characteristics of the other two.

Not all synapses are equally effective in transmitting neuronal signals: some are strong, others are weak. The effectiveness of a synapse can be altered by the signals passing through it so that synapses can more or less permanently learn from the activities in which they participate. This dependence on past history, or plasticity, acts as a memory which is possibly responsible for an important part of the human ability to remember.

The central nervous system contains several hundred different species of neurons. A casual look down the microscope will identify many morphologically different types, and even though some neurons closely resemble one another, they may well have markedly different biochemical characteristics. Different regions of the central nervous system utilize different types of neurons to process the information passing through them. Evolution has had the opportunity to make the most precise and elegant refinements to neurons and also to the architecture of the network that joins them. The neurons in the artificial networks being produced today, and which are described in detail in later chapters, are to a much higher degree of the standardized, mass-produced, "off-the-shelf" variety.

MAPS IN THE HEAD

The brain investigates its surroundings with the aid of the senses. Without the senses, the brain would probably go crazy with loneliness. If one falls asleep lying on one's arm it feels as though one has lost the whole thing, even though it is only the sense of feeling which has been lost.

A great deal of the brain's work is devoted to gaining an idea of the surrounding world. Just as we use maps to orient ourselves, the brain constructs "maps" of its surroundings and of the body's place in them.

For example, the brain has on its surface a comprehensive picture of the body's surface. This picture is not exact in scale or size, but resembles more a patchwork quilt with pieces of different sensory areas stitched together. Some regions are better represented than others. Figure 4 shows a sketch of the body in grotesquely distorted form, a homunculus, superimposed on an outline of the brain to give some impression of the amounts of information the brain receives from different parts of the body's surface.

There is a constant self-organizing struggle going on in the cerebral cortex over the exploitation of the area available. The struggle is between the different parts of the body, and those that display the highest activity win

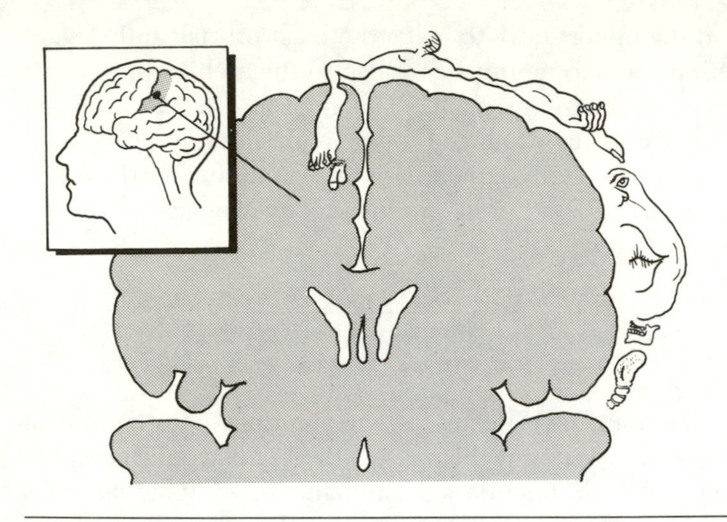

Figure 4: Homunculus mapping of the surface of the body on the cerebral cortex.

the biggest areas. Conversely, areas partly isolated by illness or trauma, are to some extent taken over by other functions.

The most demanding map in terms of resources is the one concerned with the processing of visual information. On the retina of every eye the lens imposes a two-dimensional picture of the field of view. This picture is transmitted to the vision-processing centers of the cerebral cortex at the rear of the brain. Those parts of the retinal picture which are close together, also tend to stay close together in the early vision centers of the cortex. During visual processing, shape and color of the picture are separated and these aspects of the image travel different pathways before being mixed together again and understood.

Each of the ears perceives its surroundings in one dimension in the form of sound waves which are converted to nerve signals in the inner ear. The function of the inner ear is to analyze the frequency of incoming sounds. All the way from the ear to the auditory centers of the cortex the nerves are ordered according to pitch. On the cortex there is a little tone scale with the frequency-analyzed sound signal represented as neural activity.

The chemical senses, smell and taste, are different from the others, because in their case part of reality in fact enters the sensory apparatus. When we smell rotten fish, molecules of the rotting fish actually enter our nose and react chemically with our cells. It is not strange that we are disgusted. The sense of smell also uses part of the cortex, but in this instance, as far as one knows, there is no actual map.

Altogether, we can say that the brain attempts to compile a representation of its surroundings that preserves the relationships between neighboring components. In information processing terms this is a highly economic strategy, because the brain needs to find out what parts of its surroundings relate to each other.

KNOWLEDGE

How can we know something?

It is difficult to talk about knowledge because we need knowledge to talk about it. Knowledge is a strange self-referencing concept. But it is no illusion. We all agree that knowledge exists and must have some form of reality. A great deal of business is devoted to making physical media to hold knowledge: books, gramophone records, compact discs, diskettes, films and videos.

Knowledge, like the essence of life, is easier to describe in terms of processes than in terms of properties. Knowledge can be stored, moved and processed. In this respect knowledge is just like any raw material — grain or feedstuffs for example. The difference is that grain has an immediate and unequivocal physical existence, whereas knowledge is abstract and immaterial.

The same knowledge, for example a recipe for bread, can be found in different media, *represented* in different ways. When human beings use knowledge, it is mostly done by means of *language.* A language is a common agreement on how to arrange series of elementary symbols in statements that contain meaning. The same bread recipe can be found in different books in various languages. Grain is immediately understood by its function: it can be made into bread by a Russian or French baker, even if it were grown in the United States. Grain does not need translating. Knowledge expressed in English, however, must be translated before the Russian or French baker can understand it. A language is an arbitrary representation of knowledge and it makes no difference in principle whether a recipe is printed in a book in Russian or French, or for that matter transcribed as a magnetic pattern on a diskette. In all instances, the idea is to exploit the inner degrees of freedom of a physical system to represent the statements of a language.

On a page of this book there is room for an average of 2000 characters in the typesize used. If one includes capitals, lower-case letters, numerals, spaces and punctuation marks, it means that there are approximately 75 different signs that can be made on each available space. Thus, the number of different pages which could be produced using these signs is 75^{2000}, an astronomical number, which could also be written as 1 followed by 3750 zeros. Of that horribly large number of possible pages, we have exercised the freedom of the press to produce a mere 180 for this book!

Natural languages are not perfect representations of knowledge. For example, the word *mind* exists in English but it is often not possible to find the exact equivalent of its various meanings in other languages.

Nontrivial knowledge is always structured, and structure entails a greater or lesser degree of complexity. Total order and total disorder both signify triviality. The frequently misused phrase "Everything is relative"* is

---

* In any case, Einstein never said or meant that everything was relative. On the contrary, he defined relativity very precisely, and certainly never intended to use the principle of relativity as a justification for spinelessness or equivocation of values.

trivial because it reduces all knowledge to total order. Were the words in this book to be arranged haphazardly, the knowledge they represent in their present arrangement would be reduced to a meaningless jumble.

Nontrivial knowledge is always complex and therefore has a nontrivial form; otherwise it would not be difficult to comprehend, which it usually is. Complex knowledge is difficult to acquire, because it has a profound asymmetrical structure, not easily penetrable in advance. The nonrandom structure of knowledge manifests itself in chunks or nuggets. Complicated knowledge is chock-full of nuggets.

One of man's principal activities is trying to discover internal relations in the enormous amounts of sensory inputs he is bombarded with. The brain works nonstop categorizing impressions, trying to find the pieces that will clump together. The ability to classify is a legacy of evolution. It is necessary to be able to distinguish a harmless grass snake from a poisonous adder, or a dangerous toadstool from a field mushroom. The survival of the species depends on being able to tell the difference between men and women, even if the difference can be very little. Recognition of words in sound is another subtle form of categorization, one which takes many years to perfect.

Classification of knowledge aims at masking part of the complex structure inherent in a particular piece of knowledge. Classification has meaning only in relation to the specific use of the knowledge it relates to. Can one in fact talk about structure of knowledge without giving a declaration of intent for its use? Even a recital of all the active telephone numbers in the phone book is intent-related simply because we call these numerals telephone numbers. If we did not know what they were, they would represent almost no knowledge.

Just as it is difficult to form a convincing argument for the existence of a universe independent of observers, so it is difficult to form a convincing argument for the existence of knowledge without a conscious mind to state its intended use.

When *combining* knowledge from different sources, it is not possible to determine the intent behind the combined knowledge from the intents behind its component parts. Registering the names and members of a political party can be used for the purpose of sending them a membership journal. Civil servants' names are, of course, recorded so that they can receive their salaries. However, combining the knowledge contained in those two sources could have a much less innocuous purpose, that of preventing members of certain political parties from public employment, known in Germany as *berufsverbot.*

In conventional artificial intelligence, this problem of combining knowledge has put a spoke in the wheel of many attempts to build "common sense" into computer programs. That "bartender" of the brain, who sees through the intent with combined knowledge, still has a few tricks up his sleeve that we have not yet worked out how to do.

INFORMATION

Knowledge is, as mentioned, a difficult concept to define because it is self-referencing.* One has to accept it intuitively without resolving it into other concepts.

Information, however, is better defined. The precise

* This footnote is self-referential and has no other purpose.

technical definition was given by the mathematician Claude Shannon in 1948, in connection with calculations about telephone line capacity. This definition separated knowledge from its subjective meaning and defined the informational content in terms of redundancy or degree of repetition. The informational content of a book is measured by the smallest number of simple questions one has to ask to get to know the text without reading it. Thus, the more redundant it is, the less information it contains, just as it is not particularly informative to hear the same jokes or news broadcasts repeated over and over. In this way one is led to a measure of the informational content of a particular piece of knowledge.

Shannon was only interested in knowledge which could be expressed in languages, where messages are sequences of a *finite* number of letters, so-called digital languages. The informational content of a message is high if it produces great amazement, low if little. Those messages which produce much amazement are the rare ones and vice versa. Informational content is thus related to its shock effect and is inversely proportional to the ease in anticipating it. Information is therefore a decreasing function of the messages' probability.

The simplest digital alphabet contains two letters (a one-letter alphabet would not be of much use, except to create headache). What those two letters are called is irrelevant. Morse Code, for example, uses a short and a long beep for dot and dash. The two letters are mostly written as 0 and 1, but they could equally well be F for false and T for true. When numbers are used, the messages are called binary numbers, and the digits are called *bits.**

---

* An abbreviation of binary digits.

A binary message of 8 bits could, for example, be 01101001. As there are two possibilities for each of the 8 positions it is possible to compose $2^8 = 256$ different 8-bit messages. In a situation where all messages are equally probable, one will have to ask 8 questions in the right way to reveal what the message is. Simply ask: is the first bit 0 or 1? Is the second bit 0 or 1? And so on to the eighth question. Note that the answer to these questions is simply "yes" or "no".

It is obvious that the more questions necessary to decipher the message, the more information it contains. If we decide that an information unit is the amount of knowledge provided by one "yes/no" question, then there are 8 information units in an 8-bit message. The length in bits of a binary message is a measure of its information content, when all messages of that length are equally probable. For this reason, the international information standard is also called a "bit". This definition of information has the advantage of being additive. If two independent messages are concatenated to form a third, then the resulting information is the sum of the information contained in the individual messages.

The genetic code is an example of a language with a 4-letter alphabet A, C, G, and T. These letters are abbreviations for four types of nucleotides bound together in the long DNA chains. There are about 3 billion letters in our genes, the recipe for a human being, the human *genome*. As these 4 letters can be represented with 2 bits, A = 00, C = 01, G = 10, T = 11, it means there maximally can be 6 billion bits of information in the human genome.

English has a 26-letter alphabet which can be represented by 5 bits: a = 00000, b = 00001, c = 00010, and so on. With this representation there is, in fact, room for 32 letters, 6 more than the alphabet contains, so the correct average number of bits of information per letter of the alphabet is really 4.7 instead of 5. The average information per letter in written English text is approximately 4 bits, which is somewhat less than the maximum, because the individual letters do not have the same probability of occurrence in English. It is more probable that a letter in an English text will be an `e` rather than a `k`. Furthermore many messages can be readily understood, even though they lack one or more letters. The words `averge` or `yday` can be easily understood as `average` or `yesterday` without having to pose a single "yes/no" question. Taking this into consideration, the average information per letter comes down to about 2 bits.

One can arrive at this figure both by linguistic analysis and by experiments with humans. It is of course necessary that the experimental subjects are fluent English speakers. So this way of quantifying informational content implies that there will be much more information in an English book for a Chinese who does not understand English, than there is for the average English speaking person! The Chinese cannot limit himself to the universe of possible English texts and thus has to pose many more questions to determine all the individual letters in the book. Similarly, a book with a completely random sequence of letters, typed by a chimpanzee for example, will contain much more information than any Shakespearean drama of the same length. This illustrates how the mathematical definition of information is at variance with the intuitive notion of knowledge. Knowledge seems to be contained in complex messages that are neither trivially ordered nor completely random.

STORY OF A

To store, move and process knowledge it is necessary to give it a physical representation. Normally, this problem has been taken care of by some technician, and we are only presented with the final product.

Think of a letter A as it appears on a TV screen. It represents abstract knowledge of A for whoever wrote it in the studio. Let us imagine that it is physically represented by being chalked on a blackboard. Before it can be transmitted it has to be illuminated so that it is now represented by visible light. The camera converts the visible light image to electric impulses which via amplifiers and cables are converted to electromagnetic signals that are emitted from the transmission antenna of the TV station. The A is now once more represented by electromagnetic waves similar to light, but with a different frequency and coding from before. Picked up by the receiving antenna, it is converted by the TV to electrical signals which control the electron guns in the picture tube. In the picture tube, the A is now represented by electrons in motion which are arrested by the fluorescent screen causing its atoms to light up. The A is once again represented in light which hits the eye, where it is translated to a pattern of activity in the brain's neural network corresponding to the abstract A.

All the various physical representations that A has passed through have had sufficiently many degrees of freedom to hold the "A-ness" of the letter. The TV faithfully — we are not discussing credibility of content here — transmits light and color patterns and has no knowledge of the A's "A-ness". A normal TV picture rep-

HUMAN

Figure 5: The same letter can have both "A-ness" and "H-ness". The sign also takes meaning from the context formed by the surrounding letters.

resents A graphically, not symbolically. Text TV, however, represents letters symbolically with just 7 bits per character, instead of the thousands of bits which would be required to depict an arbitrary A graphically. Thus, many hundreds of text pages can be transmitted to a TV receiver without disturbing a soap opera which is being broadcast graphically at the same time. A soap opera is by the way an excellent example of a message that requires a very large number of bits to communicate zero knowledge.

The "A-ness" of A is a strange business. In Figure 5 the word HUMAN can be read without difficulty; however, the H and the A are identical! Thus, the same symbol can have both "H-ness" and "A-ness" depending on the circumstances in which it occurs. Although people have no problem in reading the word, today's electronic scanners find it extremely difficult to distinguish between the two possible interpretations. This is because the human brain is, naturally, able to employ much wider contexts than a computer can to identify individual letters.

MEMORY

When we write a telephone number on paper, we are using the paper as a memory. Later the number can be read and used. The paper, like all forms of memory, has a limited life. In time, the writing will fade or the paper, and with it the phone number, will be lost.

In living organisms, just as in computers, it is memory which permits future behavior to depend on past history. It allows connections between events widely separated in time. For a person the separation between an act and its causes can be an entire lifetime.

Today information can be stored in many media. Before the invention of writing, man could only use the memory in his brain. Long stories were passed on orally and always preserved in a brain. This also provided a welcome development in the stories' content. The written word is much more durable simply because stone, skin, papyrus, or paper intrinsically last longer than people. In the last 5000 years man has collected an enormous extension of his memory in the libraries. Our highly organized societies could not exist without this pool of knowledge. Book burning is cultural suicide.

There is a direct relationship between the increasing complexity of the social machinery and the capacity of society's collective memory. In the last few decades we have experienced a vast extension of society's memory in the form of enormous computer databanks, the purpose of which is said to be increased service, but in reality means increased control. The remarkable thing is that in spite of all this increased control and supervision, service and individual freedom are experienced as diminishing. Only those who understand the mechanisms of control, and can learn to manipulate or get around them, have gained greater freedom.

There is probably no way of avoiding this development. For most people, the number of daily transactions with society is vastly greater now than it was a generation ago. Radio, telephone, and television have enormously increased the exchange of information. We expect to be able to draw money from our banks with a cash machine at every street corner, and this presupposes the existence of an ever active memory in the bank which almost instantly gives access to a large database.

While the banks' use of computers is unavoidable, we may well question their uses in other contexts. Via its statistical supervisory units, the State has direct access to the national socioeconomic status. Legislation has provided the State with authority to gather information without any safeguards that such information will be put to sensible use.

We are bombarded daily with economic prognoses based on such information, even though it is doubtful whether many of the economic models utilized are capable of outputting more than random numbers. Projections of the balance-of-payments' deficit and of the required number of doctors and teachers have proved to be widely inaccurate in recent years. While in the long run a legislation about merging of databases is hard to administrate, because we are good at inventing new intents for combined knowledge, a much more restrictive attitude towards the collection of data would be a better safeguard against misuse.

DNA AS MEMORY

Not all memory is man-made. Nature has its own way of storing information. In living organisms information about their form and organization is stored in the DNA molecules. The language used is similar to that of other libraries and databanks. As mentioned above, this alphabet has 4 letters, as opposed to the 26 of the English or the 2 of the binary alphabet.

Until recently, the DNA database in every cell has only been accessible for reading. Nature has never had permission to write into this database what it has learnt from experience. That the database has changed with time anyway is due to random mutations, with Nature only influencing the selection of which of the "proffered" mutations would survive. The thought that Nature should be able to make direct entries in this database is such an abomination to biologists that "Lamarckism" has become a dirty word.*

Man and his genetic engineering imply that Nature can now write in the genetic message. Lamarckism has sneaked in through the back door. We now routinely produce to order bacteria programmed for various medical and industrial purposes. In editorial terms, this technology is, however, still at the "cut and paste" level, and genetic "desktop publishing" still lies beyond the horizon.

After sex became normal practice about half a billion years ago there occurred a veritable explosion in the number of species. The ability to actively exchange genetic information made an important increase in the speed of evolution. Perhaps this sex revolution has a parallel in the current experiments in controlled gene splicing. Intelligent manipulation of genetic information

---

* Jean-Baptiste Pierre Antoine de Monet de Lamarck (1744–1829) was the father of a theory of evolution based on the idea that acquired characteristics can be inherited.

can be the next step in Nature's development of life on Earth. The mind has gained a new role in evolution. In the same way that we now find it difficult to imagine life without sex, perhaps at some future time we shall find it equally difficult to imagine that evolution could have occurred without the participation of the mind as an absolute necessity.

FUNCTIONS OF MEMORY

All real memory systems have to have three functions: 1) entry, whereby the memory is told to record particular information; 2) preservation, to hold the information intact; 3) recall, whereby the memory delivers the information on demand.

Here, we are not talking about the physical representation of information in memory, because that is immaterial to the principle — even though it would be vitally important to the engineer building a computer memory.

THE CLOAKROOM

When entering information in memory, one often receives a key or token which allows its retrieval on demand. At a railway station, you get an actual key to the luggage locker you put your case in, in a cloakroom you get a numbered ticket or token which indicates which hanger your coat is on.

Neither the key nor the token bears any direct relation

to the case or the coat. An address of this kind is an abstract symbol for the space in the memory taken up by the stored entities and is completely detached from the content.

THE LOST-PROPERTY OFFICE

In a lost-property office the procedure is necessarily quite different. If we lose a suitcase and somebody is honest enough to hand it in, we can only retrieve it by identifying it to the attendant by a partial but reasonably precise description of its contents.

In this instance, you could say the address is identical to the content. As the contents of a suitcase are usually quite varied, it is sufficient to partly describe the contents so that the attendant can identify it and agrees to release it.

Human memory functions pretty much like a lost-property office; it is content-addressed or autoassociative. Recollections are summoned by a portion of themselves. The memory associates the item to be remembered with a fragment of the recollection.

THE DIGITAL COMPUTER MEMORY

Like the cloakroom, the digital computer uses key addressing. Computer memories are composed of a very large number of binary switches, that can turn "upwards" or "downwards". There are many different technical ways of constructing them, but the important thing is that each switch corresponds to 1 bit of information. One cannot normally address every single bit in a computer, but can collect, for example, 32 bits at a time in one "word". All such words in the computer have a sequence number which is their address and thus the key to them. Unlike the cloakroom, the programmer does not receive the address with entry of the item, but must decide for himself where in the memory a particular item is going to be stored and it is his responsibility to keep account of the space available.

In most countries the government issues an identification number — social security number, *numero d'identité, numero personal* — to each citizen. This number is a key to the person and its form is a direct consequence of the way the digital computer remembers. Before the advent of the computer, a person was characterized by his or her generalia: name, address and date of birth. This information was sufficient to identify the person, because names often had a direct associative reference to the role of the person or one of his forefathers: Miller = the miller, Carpenter = the carpenter, Peterson = Peter's son, and so on. Whereas the identification number is for the computer the best key to the person, it is totally useless for ordinary people.

INFORMATION PROCESSING

Information can be transported, converted, created or discarded, all of which comes under one heading, information processing. This term is a bit of a mouthful, but then it is not describing something simple.

Information always has a physical representation even if it corresponds to abstract knowledge. Every time information is manipulated, a physical process takes place and thus information processing is subject to the constraints imposed by physical laws. Changes in physical systems are generally known as dynamics and the systems themselves as dynamical systems.

When a tune is recorded on a cassette tiny magnets are arranged in a definite pattern on a magnetic tape. With a tape recorder it is possible to change Dolly Parton's 'Apple Jack' into Mozart's 'Requiem' — that's dynamics in the fast lane.

Physical systems used for information processing always have a considerable number of degrees of freedom. The number of tiny magnets on the surface of the tape is enormous. In this case, the degrees of freedom are characterized by the sizes and directions of the magnetization of the tiny magnets. The actual number of physical degrees of freedom is always much greater than the number of degrees of freedom available for information processing. Many small magnets work together on a tape to represent one sound oscillation or a single bit.

The *state* of a physical system is known when the status of each degree of freedom is known. On the magnetic tape the state is the actual value of the size and orientation of all the small magnets.

Dynamic changes in macroscopic physical systems occur not with jerks but smoothly and continuously. Conversely, changes in digital information happen abruptly. There is nothing in between digital 0 and digital 1. Physical systems used to represent information are, therefore, most frequently made in such a way that extreme values are used to represent the digital values. The system undergoes smooth transitions between these extreme values and then pauses until the next change occurs. One often says that the system flip-flops between the two binary possibilities.

This distinction between the continuous and the discrete corresponds to the dynamics of a soccer match. Although players and ball are continually in motion, the actual position changes only when there is a score. Even when an attack is in progress, we do not say the score is 2.7 to 2 in favor of the attacking side. And when the match is over, it is only the final score which counts.

Similarly, in digital information processing systems we are only interested in changes which affect the informational content of the system. Digital information processing proceeds in jumps from one discrete state to another. Just as in chess, there is no acceptable halfway stage between moves. Once lifted, a piece must be placed, a move must be made and completed.

CALCULATION

Information processing can thus take place in dynamic systems whose internal dynamics can be utilized to alter those physical states that carry information. One way of looking at information processing is to see it as a calculation designed to produce an answer to a given question.

Digital information processing — calculation — takes place in systems — computers — which are not infinite in size. This means that the number of discrete degrees of freedom and therefore the number of information-carrying states is also finite. For example, if the computer has only 3 degrees of freedom and each can take 10 different values, then the number of states will be $10 \times 10 \times 10 = 1000$.

A digital computer can be seen as a room filled with electrical switches which are either on or off. In this case, each switch represents one information-carrying degree of freedom, and the state of the computer is specified by the positions of all the switches.

DETERMINISM

The computer computes by dynamically changing the positions of these switches. This dynamics is normally wholly deterministic, which means that when the computer is in a given state the following state is uniquely determined. It would be fatal if PCs had the imagination to do something we had not asked them to do. Even though they sometimes seem to have a will of their own, we have to remember that when something goes wrong, it is because the programmer has asked for it. If a computer becomes imaginative before the guarantee expires, your money will normally be refunded without argument!

This deterministic dynamics is definitely not a gift from Nature. All natural macroscopic systems are plagued with noise, which give them a certain measure of unpredictability. The noise can, for example, be defects in the materials from which the computer is built. In small systems it is possible to almost eliminate these defects; in large and complex systems they are unavoidable. Another source of noise stems from the thermal fluctuations, or heat motion, present in all materials. This noise can be reduced considerably by lowering the computer's temperature, and some of the latest computers are immersed in liquid nitrogen at –196 degrees centigrade.

The art of building a digital computer depends, among other things, on eliminating the effects of noise on the calculation process. This task is made considerably easier by the binary representation of information. Binary 1 can, for example, be represented in a computer by a potential of 5 volts, and binary 0 by 0 volt. In practice, these potentials will drift to either side of their ideal values, but even though the potential drops to 4.5 volts at some point, the value is still recognized as binary 1.

Preferably, there should be a substantial gap between the physical values that represent the binary numbers. When errors do occur because the real values vary from the ideal, this can, in part, be eliminated with the help of error-correcting control codes. These codes introduce a redundancy, which means that the same information is repeated in different places. In some countries, social security numbers have such a built-in redundancy. It will not do any good, for instance, to try to cheat by inventing a random number, because the authorities can check on the spot whether it is a genuine social security number or not. One has to be slightly cleverer.*

Digital computers of finite size can only offer a finite number of different calculations. Due to determinism there are just as many possible calculations as there are states. Each state can be used to start up a calculation and once started it always continues in the same way. One, in fact, only needs to know the successor to each and every state, because then one can reconstruct the whole calculation as it flows from state to state. As there are only a finite number of calculations in a computer, it is, in principle, possible to tabulate them all.

So, when they are deterministic and their calculations can be tabulated, why bother to make computers at all?

---

* In Denmark the social security number is composed of 10 numerals, for example, 250639-4673. The first six represent the birth date of the individual (day, month, year) and the last four enumerate each person born on this day. One may check whether this is a legal social security number by multiplying the number digit by digit with the fixed digits 432765-4321 in the corresponding positions. In a legal number the sum of all these products should be divisible by 11. In the above example we have thus $2 \times 4 + 5 \times 3 + 0 \times 2 + 6 \times 7 + 3 \times 6 + 9 \times 5 + 4 \times 4 + 6 \times 3 + 7 \times 2 + 3 \times 1 = 179$ into which 11 goes 16 times with 3 left over, and thus this is not a genuine Danish social security number. It is not possible to squeeze a single crown out of the otherwise very generous Danish social welfare system by slapping this number on the counter.

UNPREDICTABILITY

Real computers have such a large number of internal states that the number of possible calculations vastly exceeds the number of elementary particles in our universe (about 2^{256}). A modern PC frequently has upwards of 100 million bits of information at the user's disposal. So the total number of states is therefore of the order of $2^{100\,000\,000}$. The table of a computer's calculations simply could not be written out! There is no room for it in this otherwise rather spacious universe.

That is why we buy computers. The results of desired calculations cannot be predicted because we do not have the table. The only thing to do is to start the computer in a certain state and see where it goes. Even though the computer is dynamically deterministic, we cannot predict the result of its calculations, but only let it carry them out. The computer is nothing but a physical realization of the table.

We buy computers because they are at the same time deterministic and unpredictable or, as it is also called, *computationally irreducible*. The difference between determinism and predictability is a somewhat subtle one. The computer is deterministic because it will produce from each initial state a result which is the same whenever it is started in that state, whether in summer or winter. The computer is unpredictable because the only way of obtaining the result is to start the computer. There is no short cut to the result. No soothsayer can by crystal gazing foretell the future states of the computer. The computer can, however, make a fortune for the fortune-teller by casting horoscopes!

THE MERRY-GO-ROUND

When one views a computer in this perspective, the result of a calculation will be the entirety of states which follow from starting it in a certain initial state. In reality, there is no difference between the calculation and its result.

If any state in this sequence is repeated, we can easily reel off the rest of the sequence, because the computer is deterministic. It will go through the same cycle for ever, periodically returning to the same state again and again. When a state is repeated there is no more to be gained from that calculation. It has become a *déjà vu* and the calculation can be terminated immediately prior to the point of repetition. We indicate this by putting a parenthesis around the repeated cycle. For example if the computer has 9 states and we begin at state 5 a possible calculation could be expressed 52(46813) which is exactly the same as 52468134681346813... .

There are no infinite calculations in a digital computer. As the number of states is finite, sooner or later, one or another state must be repeated. But even in a mere PC, the number of states is so great that it could continue to run without a *déjà vu* for longer than the lifetime of our universe if only it were started up correctly.*

* Most PCs have a clock which beats once for every calculational step. It may, for example, beat 5 million times per second. The age of our universe is about 15 billion years. In units of the PC's calculation steps, it only needs 81 bits to make a calendar to span this period. An 81-bit counter in a PC could have been started at the dawn of time and would still be running without ever having repeated itself.

The length of a calculation, the number of steps before it repeats itself, is a measure of the logical depth of the calculation. If the computer is started up in a random state, it will usually not take long before the calculation starts to cycle. The computer is "hung" in a loop. Because of the unpredictability it is not possible to devise a general method of determining when a computer will begin to loop when started from a particular initial state.

To obtain long calculations it is necessary to select — to program — the initial state very, very carefully, otherwise the computer will go 'down'. Today, a huge industry is devoted solely to programming. Programmers earn their living from the construction of these initial states. Due to the enormous number of erroneous initial states, it is in fact much harder to program a computer than to find the proverbial needle in a haystack.

So sensitive is the digital computer towards the choice of initial state, that a single incorrect bit among millions of correct ones will ruin the whole thing. If digital computers are used to control machines, a jumbojet for example, one incorrectly programmed bit can cause a fatal catastrophe. The bigger the program, the more difficult it is to safeguard against the *one-bit problem*. In a program assembly as huge as that of the projected U.S. Strategic Defense Initiative (the so-called Star Wars Project) programming errors are unavoidable, which could mean that this perfect defense system might start the next war!

THE WORLD'S SMALLEST COMPUTERS

The smallest conceivable computer has a state represented by one bit only. Its logical state is either 0 or 1. For

$0 \to 0$	$0 \to 1$	$0 \to 0$	$0 \to 1$
$1 \to 1$	$1 \to 0$	$1 \to 0$	$1 \to 1$
a)	b)	c)	d)

Figure 6: The four smallest computers. Each column defines a complete computer by giving its 1-bit input-output table.

such a simple computer we are able to tabulate its calculations. The table contains two entries, one for initial state 0 and one for 1. There are exactly four different ways to make this table, so there are only four 1-bit computers, as shown in Figure 6.

Each does something different: a) in fact, does nothing, merely reproducing the input given it; b) exchanges 0 and 1, this symmetrical exchange is called a logical negation; c) and d) are very boring, because, independent of input, they produce respectively 0 and 1, and have no ability to discriminate, but merely discard the information contained in the input.

All in all, these 1-bit computers are somewhat pathological. Only b) is useful because logical negation is necessary in many calculations. It should be noted that the first two are logically reversible, in that they preserve the information given them. From the result, the initial state is unambiguously determinable when one knows which of the two first computers is being used. If one gets 1 out of computer b) it is certain that 0 was the initial state, and vice versa. Conversely, the last two are logically irreversible: if you obtain a 0 from computer c) there is no means of determining how the computer was started.

Logical reversibility and irreversibility are two important concepts for analysis of calculational processes. Initial state information is discarded and lost in irreversible computers. Computers are normally irreversible because they have to reuse a great deal of the memory space used to store intermediate results. Nobody cares about the intermediate results, but their loss makes the computer irreversible.

With one bit there are four different computers, with two bits there are 256. The number of possible computers available from n bits is $(2^n)^{2^n}$ and grows so rapidly that it is virtually impossible to enumerate them beyond the first two cases, corresponding to $n = 1$ and $n = 2$. Already with three bits the number of possible computers is $2^{24} = 16\,777\,216$.

Let us look at just three of the 256 possible 2-bit computers. Their tables are given in Figure 7.

e)	f)	g)
00 → 00	00 → 00	00 → 00
01 → 01	01 → 10	01 → 11
10 → 01	10 → 10	10 → 10
11 → 11	11 → 00	11 → 01

Figure 7: Three 2-bit computers. Each column defines a complete computer by giving its 2-bit input-output table. Input is on the left, and on the right is the result of one computational step, the output.

In e) the two right-hand columns list a calculation of two logical functions AND and OR. The first output-bit (AND) is 1 only when both the first *and* the second input-bit is 1, otherwise it is 0. Conversely, the second output-bit (OR) is 1 when the first *or* the second input-bit is 1. It is 0 only if both the input-bits are 0. Technically, we say the computer ANDs and ORs its input-bits to produce its output-bits.

In the computer f) the first output-bit calculates another logical function XOR, a shortened form of *exclusive or*. This means that the first output-bit is 1 only when *either* the first *or* the second, but not both, input-bits are 1. Otherwise it is 0. The other output-bit is not used and is always set to 0. This makes f) logically irreversible and the same is the case with e). You can determine whether a computer is irreversible simply by finding two inputs that produce the same output.

The computer g) shows a reversible form of XOR. Here, the second output-bit is utilized to say something about the input that went into the computer. In fact, it is merely a copy of the second input-bit. Together, the two output-bits provide sufficient information to determine what the input was.

All these simple computers are 'hard-wired' to their purpose and therefore have no flexibility. However, this is not peculiar to small computers only; all computers have this rigidity. The flexibility that we normally associate with the use of computers comes from the ability of a single computer to simulate whole classes of smaller computers.

For example, let us take a look at just one of the 3-bit computers of which there are 16 777 216 to choose from (see Figure 8).

The way the last bit is processed shows all the input-output relationships of the four 1-bit computers (see Figure 6). The first two bits are simply used to select which 1-bit computer one wants to simulate.

Therefore, the first two bits serve as a program for the 3-bit computer's mode of operation. Instead of buying four 1-bit computers, we can program a single 3-bit computer which therefore also appears as a flexible 1-bit computer.

$$
\begin{aligned}
000 &\rightarrow 000\\
001 &\rightarrow 001\\
010 &\rightarrow 011\\
011 &\rightarrow 010\\
100 &\rightarrow 100\\
101 &\rightarrow 100\\
110 &\rightarrow 111\\
111 &\rightarrow 111
\end{aligned}
$$

h)

Figure 8: A 3-bit computer. This computer can simulate any of the four 1-bit computers.

The 3-bit computer itself is hard-wired to simulate the four 1-bit computers. Besides this computer, there are 16 777 215 alternative ways of hard-wiring a 3-bit computer.

UNIVERSALITY

The fact that a suitable 3-bit computer can simulate any 1-bit computer is an expression of universality. Note, however, that the universality costs two bits for programming to get one bit of data processed. It becomes much worse when we try to simulate computers with many bits. A program that can simulate an arbitrary n-bit computer will require at least $n2^n$ bits. To simulate any 10-bit computer one needs a 10 240-bit program. The world's smallest computers cannot be simulated with the world's smallest computers.

FRIED EGGS AND INFORMATION

A computer cannot go backwards for the same reason that one cannot undo a fried egg! Once fried the egg can never again become a fresh, new-laid egg. You can walk backwards and forwards, up and down stairs without leaving a visible trace, but it is not possible to cook in that way. Generally, cooking is an *irreversible* process.

Irreversible processes always cause a loss of information. One becomes acutely aware of such loss when pondering just what ingredients may have gone through the mincer to produce the greasy burger one is about to swallow. It is also a meat-technological problem to determine the amount of kangaroo that is contained in "prime imported Australian ground beef", or how much pork has gone into the so-called ground beef that the Danish butcheries try to sell to the Muslim countries.

Information is lost in the same way when egg hardens during cooking. The loss of information is the reason why you cannot undo a fried egg. Once it has happened, there is not enough information in the fried egg to tell you that it was fresh and slimy before it coagulated. The reason that *you* know it is that you have seen it happen before. The old-fashioned husband who never has seen an unfried egg is unable to know from contemplation of his breakfast what eggs are like when they exit a hen.

The previous section dealt with computers that were logically either reversible or irreversible. In the irreversible computer there was no one-to-one relation between input and output. One output could be produced by several inputs. This was particularly clear in the 1-bit computer which always produced a 0, regardless of input, a computer that "forgets" all about its starting point.

A normal computer simply does not have the capacity to store information about its past states. Firstly, there is little interest in such information and, secondly, space problems would demand further extension of the computer's capacity and thus rapidly lead to financial problems. Although, because of determinism, each state has only one logical successor, in most cases it has far more than one possible logical predecessor. Just like a fried egg, the computer cannot be returned to its initial state, and, normally, one is happily ignorant of what it has been up to previously.

INFORMATION AS WASTE

Ordinary information processing consists mainly of discarding information. This is also apparent from the manner in which we instruct computers to make calculations. A typical program line in part of a large calculation might well be

```
X = X + 5
```

This line appears to be utter nonsense if you are not used to it. But the command is simply an instruction to add 5 to the value which is stored in the region of memory denoted by X. The computer must read what there is in the region X, send it through its addition unit, and write the result back to the location X. At the very instant the operation is completed, all information about the previous values of X is destroyed.

Of course, the computer could keep a log of all such transactions. If it remembered it had added 5, it could subtract it, and thus return to square one. As mentioned this luxury would, in very short order, present the owner with a serious logbook storage problem. Instead, computers compute by repeatedly chipping off information until finally the result stands out in its naked glory.

One can compare the information discarding process with what a sculptor does when he "exposes" an equestrian statue inside a block of marble. His job is "just" to *remove* the surrounding material, *not* to build the statue up from the marble. Once a small chip of marble is chopped off, it is forever lost to the statue. Just like the computer, the sculptor is not able to reverse this process.

Calculation is irreversible when several different paths lead to the same result. The logically irreversible computer has forgotten the question by the time it has produced the answer. A famous graffiti scrawled on a wayside pulpit illustrates the point: *If Jesus is the answer, what was the question*?

INFORMATION AS HEAT

How does a computer get rid of the information it discards? Do people wade around in it until the cleaners sweep it up or is there a more discreet solution?

Information loss is related to friction. An ice cube sliding about in a frictionless bowl would continue throughout eternity. Its future can be predicted and its past "postdicted". If the bowl is not frictionless, the cube will eventually wind up on the bottom. Its past is no longer postdictable. We cannot know where in the bowl it has been before it stopped.

The cube's motional energy has been transformed to heat which ends up in both the cube and the bowl. The temperature of the cube and the bowl is a tiny bit higher than before. Actually, in this example, the amount of heat is extremely small, but there are circumstances when the friction produces a great deal of heat. When a space shuttle re-enters the Earth's atmosphere, the friction is so violent that the outer surface of the spacecraft literally glows.

Information loss is always linked to heat production; it also occurs when bits in a computer are being overwritten. Heat is totally random motion in a system with many degrees of freedom. When energy is converted to heat, it does not disappear but is spread out over all degrees of freedom of the system and is therefore difficult to convert to useful work again. If the heat development causes the computer's temperature to rise it becomes harder to ensure digital precision of the machine, and it is thus necessary to get rid of the heat. Every large computer has a powerful cooling unit to maintain a constant temperature. The reliability of electronic components depends on temperature, and it does not have to increase much before it is goodbye to determinism in the way the computer operates.

So the computer gets rid of discarded information in the form of heat that is removed by the cooling system. Most of the removed heat does not in fact stem from information loss due to reuse of bits, but from all other forms of physical processes that for technological reasons are imperfect. The resistance of electric wiring to the passage of current causes friction which produces heat. It is possible to avoid this with the use of superconductors, but this necessitates the computer being maintained at a very low constant temperature, and that doesn't come cheap either. Heat production from a modern supercomputer's discarding of bits is of the order of 10^{-14} kilowatts while its cooler unit has a capacity of about 100 kilowatts.

The actual energy expenditure for overwriting bits is, however, rapidly decreasing with time. Over the last 50 years it has decreased by a factor of 100 billion! If this development continues for another 20–30 years, one will reach a point where the thermodynamics of computational processes will play a major role. The only way to further diminish the energy expenditure will then be through the construction of logically reversible computers.

Reliability is essential for computers, and there is nothing more depressing than an unreliable one. A computer that cannot be counted on is soon out of the door either to be repaired or to find eternal rest at the junkyard.

Both logically reversible and irreversible computers use energy to maintain the dynamic progress on the right track from the initial to the final state. The physical representation of a logical state, a 5 volt potential for example, must not drift too much before it has to be put back on the track. Any errors in the computer's physical representation of digital information arising from noise or minor faults must be eliminated as soon as possible. The error must be forgotten.

If an error is large enough to affect the logical state, it can only be corrected if there is redundant information built into the states of the computer. This type of redundant information is reminiscent of the control codes in a social security number. To correct the error, it is necessary to overwrite the incorrect information and such correction demands expenditure of energy.

Such readjustments or corrections must, as with elderly buildings, be made before it is too late to recognize the original state or condition. Once it is too late, one has to begin again, either with a new calculation or a new house. If, however, the mistake is picked up and corrected in the course of the calculation, the correct result can still be attained, even if the computer functions in a noisy environment.

Energy consumption in a digital computer is in principle associated with the destruction of information, not its creation. A calculation in a computer never creates information, but only throws it away.

COMPUTER ARCHITECTURE

Business procedures in Russian butcher shops are based on the art of queuing at a very advanced level. Customers are served one at a time and always in the order in which they reach the counter. If a particular customer wants something from a distant shelf, or orders a particularly difficult cut of meat, the entire queue must wait patiently. The butcher will fetch the meat from the cold store, dress the correct cut on the block, wrap it at the counter, work out the price and only then shout: *Next!*

The single-file queue system certainly facilitates the control over the activities in the butcher shop. All activity is concentrated at a single point. The butcher always knows exactly what he has in stock because nobody else has access to it. The business functions after one of Lenin's axioms: *Trust is good, but control is better.*

In a modern supermarket, where customers move about independently serving themselves, the process is just the opposite. All customers at the same time select the goods they want. If the health authorities had not prevented it, they could just as well fetch their own sides of meat and joint them out for themselves. The supermarket control problem, however, is much more difficult than in the Russian butcher shop. If the supermarket did not invest at least a certain trust in the customers it could not function and would be back at the level of the single-file queue

Figure 9: A Russian butcher shop. (Photo: Politikens Pressefoto).

system. The time saved by self-service would be lost again in the control measures.

The supermarket's trust, however, extends only as far as the checkout, where the queues reappear because each customer and purchase must be individually checked. Of course, this could be avoided if each customer were trusted to total the cost of their purchases, ring them up and put the money in the till. This is rather like neighborhood parties with a communal beer pool where everyone is on trust to put the money in for each beer they grab. In practice, however, this system usually proves less than perfect.

The problems of management is a science in itself, particularly in giant organizations like the mail service or the armed forces. When the activities of large numbers of people have to be organized, the traditional answer has always been control, control and more control. In the Western democracies, with the increasing recognition of the individual's right to self-determination, control has diminished in proportion to the diminishing ability to impose sanctions. Today, it has reached the stage where in some countries it is even prohibited by law to spank naughty children.

In the Prussian Army and its 20th century successors control was absolute. All activity was determined by explicit orders from above. When something went wrong or a link in the chain of command was broken, problems were inevitable. The higher up the chain the madness occurred, the worse was the resulting foul-up. We have all heard war criminals, sometimes heads of state, excuse themselves with the time-worn formula: "I was only obeying orders".

Democracies are inevitably trapped in the unavoidable dilemma posed by the conflict between the rights of the individual and the established and necessary organization of society. As nobody wants his individual rights diminished, the only way out of the dilemma is a less hierarchic organization of society. However, apart from a few concessions towards decentralization, governmental reaction is usually just the opposite: control, control and more control.

Without sanctions — spanking, the stocks, debtors' prison, execution — it is pointless to try to maintain rigid lines of control. The ever-increasing complexity of society means that the controllers' controllers have themselves to be controlled and so on in an endless chain which eventually closes on itself. The logical end result would be that almost the entire mechanism of state becomes an organ of control, spreading like some vast bureaucratic tumor until it has consumed the whole population.

VON NEUMANN'S BOTTLENECK

The ordinary serial computer is exactly like the Russian butcher shop. The "butcher" is the central processor, the stock is the data to be processed, and the customers are the instructions to be carried out. At any given time, the central processor is only dealing with one instruction. The shop assistant in the computer, the butcher, is kept exceedingly busy because goods are being constantly fetched from stock and, thus, a large part of the assistant's time is spent dashing backwards and forwards. Conversely, in the stock room all is peaceful, because there is activity on only one "shelf" at any given moment.

This fundamental problem of moving data between storage and processor creates a *bottleneck* which crucially limits the computer's performance. In 1978, the computer scientist John Backus named this bottleneck after one of the fathers of the modern computer, the mathematician John von Neumann.

A look into a modern PC demonstrates the problem. The central processor is normally encapsulated in a single *chip* with a number of legs sticking out of it. These legs or pins are plugged into a socket soldered to connections with other parts of the computer. Sometimes, the central processor may comprise several chips, but this only

Figure 10: von Neumann's bottleneck (Photo: Søren Hartvig)

makes the communication problem worse. In other locations an army of *memory chips* is arranged in rank and file.

Physically, von Neumann's bottleneck is the pins sticking down from the central processor chip into the socket. All communication with the outside world takes place through these pins; they are the door to both the butcher's stockroom and to the street outside. Even though there can be as many as a hundred pin connections per chip it is still an incredibly narrow channel. Some pins are used for power supply and others for the clock used to synchronize the processor's activity, but most of the pins are reserved for communication with the data storage.

The Russian butcher shop analogy is not quite perfect,

because the customers' orders — the computer's instructions — are also contained in the storage. Health and hygiene regulations forbid this sort of mixing of customers and food in a real butcher's store — even in Russia — and the very idea of butcher's shops itself might be lost if customers lived in the cold store.

SERIAL ARCHITECTURE

There were two very different reasons why, in the computer's infancy in the 1950s, this two-part design was chosen, with the data storage and the processing function kept separate. The first was entirely financial and technological, while the other reason was more psychological.

The whole idea of the computer is to make a machine which is capable of processing certain kinds of information faster than we can do it ourselves. In the 1950s, the only fast components available were relatively expensive radio tubes. In order to get any value for the money invested it was necessary to keep the rather modest number of such components one could afford extremely busy.

Conversely, a computer was of no particular advantage unless it had a large number of logical degrees of freedom and this could only be achieved with the aid of slow and cheap components. It was at this juncture that the division occurred between the fast, effective and costly central processor and the slow, cheap memory banks. This separation met the double requirement for both size and speed.

While those were the hard facts in 1950, they do not apply today. If we look inside a modern PC, there is little visible difference between the processor chip and the memory chips, apart from their size. If you were to open them up, you would need a microscope to see anything. The network of wires in a processor chip resembles the convolutions of Calcutta's street plan, while that of the storage chip has all the regularity of Manhattan. This difference in microscopic structure reflects the difference in complexity of the tasks that the two chips perform. The basic substratum for the wiring is in both cases silicon, which is the main substance in ordinary sand. None of the components in a chip costs much. Even the production process now requires little or no human intervention and is therefore also relatively inexpensive. The only real expense is the cost of the design process, but this is offset by producing the chips in enormous quantities. Even though the processor still is more expensive than the memory, their prices today are becoming comparable. Although we can now without any difficulty produce storage chips with a lot more built-in processing capacity, the major problem is to decide what processes they should carry out.

The second reason for splitting computers into an active processor and an extensive inactive data storage is that it is difficult for us to think along other lines. Man's own conscious thinking is *serial* or sequential in time. It is impossible for us to read seven books simultaneously. Our very use of language is sequential in time. Words follow each other one by one and, as a general rule, it is difficult to follow a discussion when several people talk at the same time, a technique much used by professional politicians to sabotage their opponent's contact with an audience. To obviate this form of skullduggery in TV debates, it has become necessary to have only one microphone open at a time.

The serial nature of conscious human thought is clearly demonstrated by the way we learn to perform the simplest calculations. When two multi-digit numbers have to be added, we break the problem into smaller parts and carry them out one after the other. We start on the right, adding the least important digits and work our way leftwards in ascending order of magnitude with appropriate markings for carries.

The von Neumann's bottleneck is in fact two-necked because it is not only a physical bottleneck in terms of data transmission, but also a mental bottleneck which makes it difficult to think in terms of another calculational scheme which does not operate within what is called "the one-thing-at-a-time paradigm".

PARALLEL ARCHITECTURE

The supermarket is a somewhat pathological analogy for parallel processing because the exact order of the sequence of "instructions" carried out by individual customers is more or less irrelevant. It makes no difference whether the customer collects coffee before cream or milk before meat. Nor does it matter how the various customer's shopping lists vary, unless they all want to use the coffee grinder at once. During sales one may witness the more general case where the sequence of operations becomes relevant, because many customers want the same item at the same time.

The supermarket is an example of almost *asynchronous, uncoupled parallel* processing. A symphony orchestra, however, demonstrates highly synchronized parallel processing. It is still uncoupled, inasmuch as each musician plays his own instrument. Each has his own "program" of notes and "simply" has to follow it and the movements of the conductor's baton. This uncoupling or disconnection makes it possible to record both opera and rock music without the different artists being present at the same time.

In more general cases of parallel processing the order and nature of the instructions are of much greater significance. It demands a great deal of vigilant management when several simultaneously working processors have to be coordinated.

On a newspaper, there are many independent functions which can be performed simultaneously. There are sports reporters, reviewers, gossip columnists and political commentators all struggling with their own specialities. This independence is only limited by the *deadline* which initiates all the subsequent processes of printing, dispatch and distribution. While certain subprocesses are carried out independently of each other, there are set times by which these subprocesses must be completed and new parallel subprocesses set in motion in the long flow of coordinated activity which is the production of a daily paper.

There are many different ways of speeding up production by parallelization of subprocesses. Parallelism can be separated into classes by the way the subprocesses relate to each other. Nearly all methods implemented in computers find analogous routines in daily life.

MASS BAPTISM

A certain suburban vicar was very popular in his parish. Many churchgoers wanted to submit themselves to *his* expounding of the Holy Scripture. This gave rise to some serious queuing problems in his church on Sunday mornings when too many people wanted to have their children baptized. The vicar solved this problem in a manner which in "computerese" would be called *vector processing.*

The first child was read the complete ritual, right down to the fine print, but the following children only got the headlines with the added rider that the details spoken over the first child applied equally to all the rest.

This is, of course, not true parallel processing, but only the reuse of the same instruction with different data. Because the church in question only had one font, all the children had to pass it serially, but time was saved on the initialization procedure. Only the first child received the full initialization while that of the subsequent children was abbreviated to the shortest possible. By having the parents queuing ready, child in arms, the vicar also eliminated the usual time wasted in getting up from their pews, readying the child, moving down the aisles, and what not.

The modern computers marketed under the name of *supercomputers* draw their strength from vector processing. With these computers it is only necessary to retrieve certain commands once from storage; after that they will work through long "queues" of data called vectors. In this way the constant retrieval of new commands from storage is avoided. Each time it is necessary to handle individual data elements in the vectors, for example the multiplication of numbers from two sequences, only one instruction is fetched, just as a true chain smoker uses only one match in the morning to light the first cigarette. Many scientific calculations can readily be *vectorized* and, thus, utilize the special resources of this type of computer. A well-vectorized program may improve calculation speed by a factor of 100 compared with nonvectorized calculation.

THE CONSTRUCTION BOSS'S TRICK

Big-time constructors do not actually do any building themselves, they use subcontractors. As usual, the whole idea is to get someone else to do the actual work, screw them on the price and pocket the difference.

Subcontractors are specialists in concrete work, electrical fitting, plumbing, brickwork, and carpentry, but none of them is actually able to build a complete house. The construction boss's job is to coordinate the work produced by all these specialists, usually a difficult task in practice, as subcontractors tend to have scant respect for each other's needs. The construction boss is the serial element — the consciousness — in the building process. The boss controls which processes can be going on simultaneously. Plumbers and electricians can work at the same time, but not before the concreting is finished.

Even though a computer processor is called serial, in fact it nearly always carries out a small amount of parallel information processing of this type. The processor contains "specialists" that can, for example, add, multiply or carry out various different logical operations. When the output from one part-operation is not needed for the input of the next, they can occur simultaneously. In the calculation $A = B + C - B \times C$ it is necessary to calculate both the sum and the product of the same two numbers before the results can be subtracted. These two stages or part-operations are entirely independent of each other and can be carried out in different sections of the processor if it is made for that. However, the subtraction cannot begin until the results of the addition and multiplication are ready.

This simple parallelization can take place automatically because the processor is operating on very few data objects known as *registers*. It is therefore easy to assess the dependence of successive instructions, and this form of parallelism has, therefore, been a fixed constituent of most processor architectures in the last 20 years.

CHIEFS AND INDIANS

The construction boss's trick is not so difficult to understand. The operations to be carried out have a direct relationship with the available specialists: electricity is installed by electricians, piping by plumbers, brickwork by bricklayers. Problems only arise when a building project is so extensive that no single construction boss can handle it.

When a project is very large, the type of management changes. If the project requires not 3 electricians but 450, their management and coordination is a whole new ball game. It becomes necessary to introduce one or more levels of foremen, a kind of hidden units, that never touch a screwdriver.

Hierarchical elements seem to emerge naturally in all human systems of great complexity. This is necessitated by the individual's limited ability to process information. Their communication channels have too little capacity or bandwidth. A chief who could conduct twenty thousand discussions simultaneously would, at least theoretically, be capable of running such a huge organization without middle management or supervisors. But real chiefs are human and are thus restricted by the physical limitations of two-way communication. People can only interact with one person at once and there are only 24 hours in a day.

The real task of a hierarchy is to discard information in a smart way. The whole idea of efficient management or gathering of intelligence is to select and retain only that information which is necessary for control at a higher level and layer for layer to discard the welter of unnecessary details.

A similar function occurs in computers. If a computer has several processors they are normally ranked in a well-defined manner. One processor is the "boss" for the whole system, distributing and directing the work. Under this layer one finds the processors the ordinary user can program. At the lowest level there are a number of slave processors that handle elementary input/output functions. However, unlike in human systems, there is no promotion in computers.

Even though a computer in a bank may contain several processors, as a rule there is no collaboration between the processors on individual tasks. As most of the tasks, paying in and paying out for example, are relatively small and largely independent of each other, every processor can carry out its function independently. The "big chief" processor directs and makes sure that the other processors are kept as busy as possible while at the same time ensuring that the functions laid down by the program are carried out in the correct order. In this way the computer as a whole can speed up its operation when, as is frequently the case in practice, there are many tasks to perform.

Conversely, the speed cannot be increased if there is only a single task to perform, however large it may be. To utilize all the processors it is thus necessary to divide the job into smaller parts with clear interdependence. This problem is nearly unsolvable for the conventional computer and if parallelism is to be exploited, it requires hard programming to implement the break-up into subproblems of each new task.

THE CONNECTION MACHINE©

If there are too many processors they cannot be made to work together in a highly hierarchical system. The hierarchical structure makes it difficult to construct an effective communication scheme that will fit many different computational problems.

Everything goes completely crazy when the number of processors exceeds all bounds, a situation called *massively parallel*. The bandwidth of the highest level is so narrow compared with the massive flow of information taking place at the lowest, that the information reaching the top can bear very little relation to what is actually happening.

If you want a message delivered fast you need good *connections.* If you do not have the right connections or contacts then the message will have to go by a roundabout route which always takes longer. In computers with a massively parallel communications requirement it is always necessary to take the spatial distance between communicating parts into consideration. Processors which are in frequent reciprocal contact should be sited close to each other.

Very little is said about connections in conventional computer architecture. It is more or less taken for granted that those parts which have to communicate will be able to do so. The situation is entirely different in the modern supercomputer, particularly the massively parallel computers. Instead of being a trivial item, the *electrical wire* has in fact turned out to be the most problematic and costly component.

The human brain has connections everywhere and most of the brain's weight and volume is occupied by its wiring. Trying to imitate the brain's architecture, traditional computer science comes up hard against physical reality. The wires, previously regarded as insignificant, have now become the main issue.

When there are as many processors in a computer as there are grains of sand in a cubic meter, their spatial arrangement can no longer be ignored. Architecturally, this means that physical considerations become of essential significance. Processors with a high need for reciprocal communication should be subject to a "gravitational pull" which will bring them closer together. In this way, it is possible to establish compact masses of strongly communicating units, like those in the human brain.

In 1985 a computer was constructed following these architectural guidelines. Daniel Hillis, its designer, called it the *Connection Machine*© and it is marketed by the *Thinking Machines Corporation*. A large version of this computer contains 65 536 processors coupled in a hypercubic net which has as many edges as a cube would have in twelve-dimensional space. Such a cube would have 4096 corners, each occupied by a single chip holding 16 microprocessors. Each corner is connected to 12 neighboring corners and as each connection handles a two-way communication, the net has a total of $(4096 \times 12)/2 = 24\,576$ connections. Even though this network with 12 connections per corner has a high degree of connectivity it still bears no comparison with that of the brain with at least 1000 connections per neuron, which in this computer would mean at least 4 000 000 connections!

If two arbitrary corners are to communicate, the message can reach its destination having being "forwarded" at the most 11 times. There are, however, $12! = 12 \times 11 \times 10 \ldots 2 \times 1 = 479\,001\,600$ alternative routes to choose from. Although processors are soldered into position and cannot move physically, with appropriate programming they can formally move by changing addresses. This can significantly reduce the mean communication time between two processors.

Thus, communication can serve as a gravitational pull linking processors together in substructures. While in the von Neumann computer the attempt is always to optimize an algorithm to fit the computer's architecture, vector architecture for example, with the *Connection Machine*© it is just the opposite, it is possible to adapt the computer's architecture to the calculation. This is a radical innovation the ramifications of which have yet to be fully understood.

The machine is not designed such that every processor can run its own private program. All processors in the machine execute simultaneously the same instructions, but each processor operates on different data elements. Each processor also has a flag which may be used for vetoing instruction execution. This makes it conceptually easy to grasp what is going at any particular moment, and therefore also relatively easy to program the machine. In this respect it resembles a vector machine, but such a machine contains only a few processors. The multitude of processors in the *Connection Machine*© makes it orders of magnitude faster than any vector machine on problems suited for this architecture.

Some of the uses of the *Connection Machine*© have been to simulate the behavior of systems with local interactions such as flow patterns around wing profiles, to reconstruct three-dimensional images, and to sort large amounts of data.

In spite of the fact that this architecture in principle makes possible a very high level of interprocessor communication, it is still far below that which is needed for the most common artificial neural networks. The connectivity of natural as well as artificial neural networks is simply too high. If one wishes to construct fast artificial neural networks, simulated connections are not good enough. The connections must be present in hardware.

BINARY NEURAL NETWORKS

Human consciousness is, by us at least, usually judged to be our most important mental component, but is, in fact, probably not a primary requirement for the basic characteristics of intelligent behavior. A consciousness of the symbol manipulating type evolved very late in the history of life here on Earth, and many animals survive perfectly without ever giving the impression that they are conscious.

In the biological world, what is significant is the speed with which calculations can be performed. When the tiger springs, it is vital to run immediately, preferably in the right direction. As soon as the tiger is in the air, it is only subject to the force of gravity — plus air resistance — and will follow a ballistic orbit. It is therefore possible to calculate where it will land. We are not, however, consciously calculating when we take to our heels. We just have a reasonably good idea of where the ballistic tiger is likely to land. But this involves, in fact, a lightning calculation that must be finished, before the reaction can take place.

If that type of conversion of a visual impression into a decision were to be undertaken by a serial computer we would have been eaten before the result was ready. Even making use of the fastest electronic components available it is impossible to compete with the speed of the brain.

The human brain is so powerful that it will never be possible to simulate it serially. However sophisticated, the hierarchical architecture of the serial computer is hopelessly unsuited to undertake the type of calculations which are the basis of human intelligence. Even though we probably should not be forced to make these calculations in exactly the same way in a computer as in the brain, it is quite clear that another principle of calculation — another paradigm — must enter the picture.

The brain's massively parallel neural network is an existence proof that intelligent systems may be created from parallel-working and, in the case of the brain, even slow components. Nature has at least succeeded in doing this. It is therefore tempting to imitate the brain's distributed method of computing by constructing artificial neural networks.

It is far from the first time that man has attempted to steal a trick from Nature. Artificial flight has been a reality for almost a hundred years, though in this case, of course, success was not achieved by directly copying Nature. All who attempted to make a flying machine with flapping wings either became common laughing stock or fell down, like Icarus and his father Daedalus, when they left Crete. The artificial flapping-wing principle has never later achieved much success.

The evolutionary construction processes of Nature can be forced to follow paths that *we* would never wish to take, and conversely we can pursue courses which would probably never be accessible to Nature. Without the discovery of the propeller, flapping wings might well have continued to be the road to mechanical flight and, equally, had Nature hit on the propeller, birds would have looked quite funny. Similarly, Nature never found an evolutionary niche for the wheel because there are no natural highways. Although animals must be off-roaders they have not been equipped with four wheels.

When attempting to copy Nature we are free to change and modify. Nearly all aircraft have, like birds, wings and tails, but use propellers or turbines, instead of wings, as their motive source. Aircraft designers learnt from Nature that flaps and rudder surfaces are necessary to control flight. Beaks, however, proved unnecessary and were discarded. Claws were replaced by wheels, the eyes and brain are the pilot's, and fuel tanks have been substituted for the stomach.

We have the same freedom to change and modify the design of neural networks. About the eventual limitations of this process we know just as little as the first aircraft designers. These boundaries can only be determined empirically and just as many of these experiments will fail and be ridiculous as were those with early aircraft.

The first artificial neural networks will also be as unbelievably primitive as the Wright brothers' flying machine. There is only little resemblance between their Flyer and today's space shuttle that takes people unscathed in and out of the Earth's atmosphere. The important thing was that the Wright brothers took the first step in the right direction.

It is probably a lot more difficult to take a convincing first step in the direction of a man-made neural network with intelligent capacity. There are in nature no small intelligent networks from which we presumably can deduce that there is a certain critical size, probably rather large, before a neural net becomes intelligent. Even the neural network of a fruitfly is beyond question more intelligent than an F-16 fighter which is currently billed

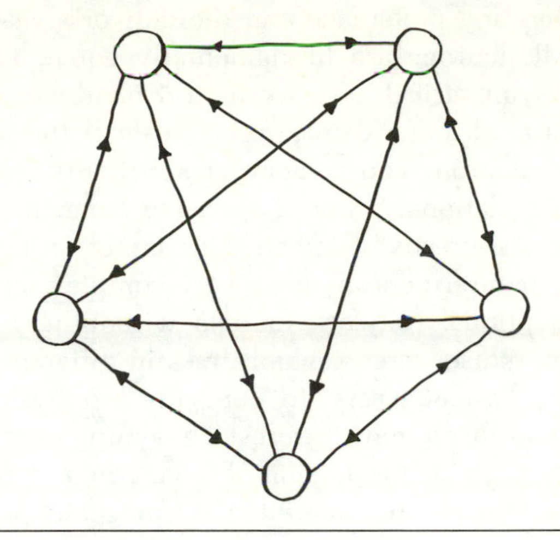

Figure 11: The simplest network with 5 neurons. Connections between them are two-way, thus giving 10 connections.

as "software wrapped in sheet metal".

THE SYMMETRICAL NETWORK

In a network it is, above all, the connections that matter. At the network nodes sit units, artificial neurons, that make decisions on the basis of information received through the connections, which in turn serve as communication channels between the nodes.

In the simplest neural network all possible connections are present, and it therefore possesses maximal symmetry. No neuron is being "cheated" because each has equal possibility of communicating with all the others. No neuron has any preference and the network has no "corners". All neurons are fundamentally equal. Figure 11 shows a symmetrical network with 5 neurons.

In this model one disregards almost all the details of biological neurons and concentrates virtually exclusively on the calculational properties arising from the network structure. As already stated in the chapter on the brain's wetware, natural neurons have very complex mathematical descriptions. Network models with binary neurons have been rediscovered several times in different variants during the last 40 years. In 1982 the biophysicist John Hopfield aroused new interest in symmetrical neural networks by giving his version of the model a form related to physical models of materials with magnetic properties.

THE ACTIVITY OF ARTIFICIAL NEURONS

To arrive at a clear mathematical description, one selects and simplifies the characteristics of natural neurons, in particular the electrochemical *activity level.* This is the frequency with which neural impulses are emitted. This level of activity determines, as in natural neurons, the degree to which a particular neuron can affect the other neurons it is connected to. In a symmetrical network this means, as mentioned, all the other neurons.

Natural neurons tend to have either a low or a high activity level, seldom an intermediate one. This feature is idealized in the symmetrical network so that the artificial neurons have only two levels of activity, either *active* or *inactive.* Neurons are also said to either *fire* or not. Therefore, the definition of the whole network's *activa-*

tion state or *firing pattern* consists of listing the activities of all the neurons. One could imagine a firing pattern as a snapshot of a lightboard where the individual bulbs are either on or off.

If we denote the active state by the number 1 and the inactive by 0, then the complete firing pattern will be a binary number with as many digits as there are neurons in the network. If, as in Figure 11, there are 5 neurons in the net, a possible firing pattern at a given moment might be 01101 or 10110. In this instance there are $2^5 = 32$ possible firing patterns.

In this respect there is no difference between the definition of the symmetrical network's firing pattern and the logical state of a digital computer. They are exactly the same. The difference between them first emerges when we consider the dynamic development of the two systems. This is analogous to the Russian butcher shop and the meat section of a (Russian) supermarket. They have the same goods on the shelves and thus the same state description, but the difference arises in their dynamics, which concerns the way the goods leave the shelves.

DYNAMICS

Natural neurons communicate via electrochemical contacts called *synapses*. Artificial neurons in a neural network similarly affect each other's activity levels via artificial synapses whose strength determines the degree of *influence* a particular neuron has on another. If the artificial synapse has a high strength then the neurons connecting through this synapse will have a considerable influence on each other and if the strength is low, the influence will be small.

It is the strengths of these artificial synapses which together are responsible for the dynamics of the network. The artificial neurons only affect each other when they are active. In a symmetrical neural network the neurons interact symmetrically, which means that the influence any given active neuron has on another is exactly the same as that which the other has on the first when *it* is active.

From time to time the neurons decide whether they shall be active or inactive, a decision based on the collective influence — the "collegial pressure" — imposed on a neuron immediately before the decision is taken. The collective influence on a given neuron is calculated by adding the individual influences from each of the neurons to which it is connected. When the collective influence exceeds a certain limit the neuron goes active, otherwise it goes inactive. Neurons are not identical, because each has a unique pressure threshold which must be reached before it fires, and a unique set of synaptic strengths.

In spite of being independent decision makers neurons are, in a well-defined fashion, forced to react to the collective influences to which they are exposed. Neurons cannot elect to remain inactive if the collective pressure exceeds the threshold or go active if the pressure is too low. The binary neurons behave as a policeman who will put you behind bars only if you are under so much influence that the alcohol content in your blood exceeds a certain threshold.

Artificial neurons, of course, do not have a "free will" corresponding to the one we think we have, and this also applies to the natural neurons in our brains. The free will that we experience *may* be an illusion. But our *will* — free or not — must be a collective effect at the macroscopic level of the deterministic interactions of many "obedient" neurons; exactly in the same way as life is a collective effect of the interactions of many dead atoms.

The influence that one neuron has on another can either strengthen or weaken its desire to fire. If the synaptic strength between two neurons is positive — or excitatory — then the influence will contribute to the recipient neuron crossing its pressure threshold. Equally, a negative — or inhibitory — synaptic strength will reverse the effect, inducing the recipient to inactivity.

Interaction between neurons is based on influence rather than control. When a neuron attempts to influence another to activity via a positive synapse, it is not certain that the recipient will be activated. The converse effect also applies, even if the synapse is negative, inactivation is not certain. The other neurons also have a role to play. A connection which perseveres in trying to influence a neuron in a given direction without success, is said to be *frustrated*.

THE FRUSTRATED TRIANGLE

Interactions between people are also dogged by frustration. We can, from time to time, be forced to adopt attitudes in conflict with those of our friends. You can't please everyone all the time. Sometimes it just cannot be helped.

Frustration is also obviously inherent in the interaction

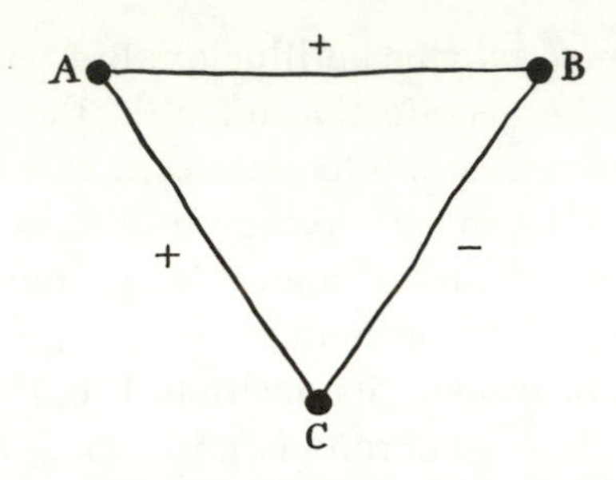

Figure 12: The frustrated triangle. A and B have a tendency to agree, so do A and C, the frustration arises from B and C's tendency to disagree. Total harmony is impossible.

between political parties. Two polar parties, for example, could well both be courting the same centrist party, and yet refuse to be in the same room with each other. Result: frustration. In a neural network there can be any number of frustrated triangles. The aim of the dynamic development of a symmetrical neural network is similar to that of the parliamentary process: to gradually reduce the amount of frustration.

If a symmetrical neural network has a firing pattern where not all the neurons conform to the collective pressure, the network will be in an unstable state of frustration. This will be lessened if just one neuron changes its state and conforms to the collective pressure exerted on it. Should all the neurons gradually thus conform, then in the end the frustration level will be reduced to a stable minimum.

When a stable minimum is attained, all neurons balance each other and none will change from the active to the inactive level or vice versa. The symmetrical network, like the digital computer, tends towards a stable state which can be seen as the result of a calculation. How the network is induced to perform a particular calculation with a certain purpose will be described in the following.

NONLINEARITY

Each neuron is an independent decision maker. In this model the neurons are deterministic and there is no hesitation about whether the state will be active or inactive. The sudden crossing of the threshold which is the basis of the decision's clarity is an expression of *nonlinearity*.

A truly linear neuron may respond with a cautious "Perhaps..." rather than a clear-cut "Yes" or "No". One may surmise that the microscopic nonlinearity which is also a characteristic of natural neurons is the reason for our ability to take clear-cut decisions at the macroscopic, cognitive level.

IN OR OUT OF STEP

In the ordinary computer there is complete control over what occurs. There is, in fact, a kind of metronome to ensure that all events take place in the correct order. The conductor of a symphony orchestra is its metronome and guarantees with his authority that the musicians play at the required tempo. If his control falters, the music falls apart as can be experienced in Federico Fellini's *Orchestra Rehearsal.*

In contrast, the neurons of a symmetrical network are independent. The network is an asynchronous computer where the individual neurons usually are out of step. Neurons take decisions incessantly and alter their states according to the pressure they experience. The network's dynamics is *robust* against individual neurons getting out of step. Such robustness is a biological necessity for the brain because the dynamics of its wetware does not have digital precision.

The same robustness is of critical importance for the construction of artificial networks from electronic components. When computer processors, even as simple as artificial neurons, are required to cooperate in conventional terms, the synchronization problem grows almost uncontrollably as the size of the system increases. In very large and extended systems it is a practical impossibility. The orchestra *101 Strings* sounds awful only for the reason that there is a delay in the sound coming from the various sections of the orchestra.

The computer networks that airlines use for ticket bookings can never be perfectly synchronized, because of their widely different geographical locations. Because of poor network connections information can take so long in transit that the same seat can be, and frequently is, sold several times over. As *overbooking* has never been a high priority problem for the airlines, they have "solved" the insoluble synchronization problem by doing nothing about it, apart from giving you a free ticket if you agree to be bumped off the flight!

THE WISDOM OF THE CANE

In the good old days, back there in the little thatched schoolhouse, the village schoolmaster kept a pretty tight grip on things. His cane encouraged the pupils to think along correct lines so that their limited school days would be used to best advantage. It is an educative method no longer used today, partly because it is not particularly effective and partly because it is in conflict with the elementary human rights of children. We have changed to a much less strictly controlled mode of learning, hoping thereby to make the childrens' minds, as opposed to their backsides, better proof against the changing vicissitudes of our times. Psalms learned by heart, Latin verbs and Euclidean theorems are of little use in a world where ingenuity and imagination are the criteria for survival.

The artificial neural network will first become interesting as a computer when it is capable of carrying out well-defined calculational tasks. In the ordinary computer such tasks are accomplished by *programming* which ensures brutally rigid control of every bit and part of the machine. Because of the neurons' autonomy, it is not possible to use such methods in a neural network and one is forced to use more pedagogical means. It is, in fact, this very lack of control which creates the possibility for the neural network to perform calculational tasks no conventional programmer or AI freak could program his way out of.

In a neural network, only *influence* is exchanged between the individual neurons. If a neural network is to solve a particular problem, an *external pressure* must be brought to bear on the complete network by the surroundings. This pressure will gradually steer the network's response in the right direction so that finally it will be capable of solving the problem. The network has been *trained.*

There is some analogy to this idea in the AIDS campaigns which some countries have run. Instead of attempting control through dictatorial legislation, the idea has been to exert influence and pressure through a massive media blitz. Tax legislation is an equally good example of just the opposite, whereby the excessive use of prohibition as control makes it legal to find all the loopholes and short cuts which must necessarily exist in such a complicated system. In some countries the AIDS campaigns have been oriented towards making the individual act responsibly instead of leaving matters to higher authority. One may wonder what a similar campaign about the payment of taxes might produce!

ASSOCIATIVE MEMORY

Human memory is based on associations with the memories it contains. Just a snatch of a well-known tune is enough to bring the whole thing back to mind. A forgotten joke is suddenly completely remembered when the next-door neighbor starts to tell it again. This type of memory has previously been termed content-addressable, which means that one small part of the particular memory is linked — associated — with the rest.

A symmetrical binary network can be trained to function as an associative memory. The memories which the network can contain are neural activity patterns which in

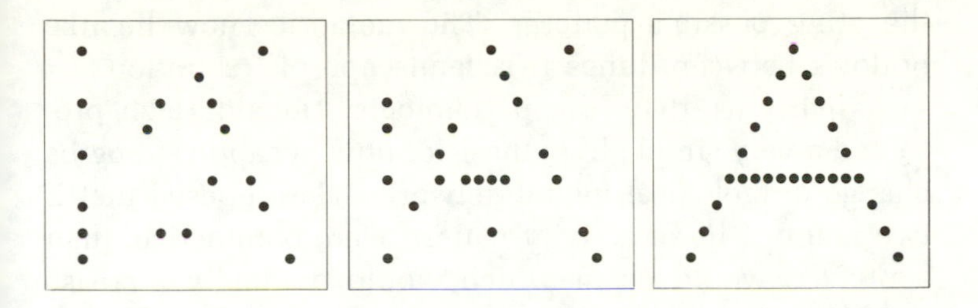

Figure 13: Dynamic development in a 200-neuron network. The start configuration is an incomplete letter which the network's dynamics restores to its complete form.

a binary system are a series of zeros and ones. Input to the network is a neural firing pattern representing the lump of information to be retained. This firing pattern could, for example, be part of a previously learned letter such as that shown in Figure 13. This fragment of the letter triggers an unstable activity pattern in the network which gradually is altered by the network's dynamics. Collectively and in parallel the net's neurons begin a struggle that will in the end restore a stable state with minimum frustration, the memorized letter.

The strengths of the artificial synapses govern which activity patterns are stable and also determine what the model is capable of remembering. These strengths must therefore be established during training.

Those patterns of activity that the network should be able to remember must be the least frustrated because such patterns are the only stable ones. Therefore, training consists of feeding the net with the desired memories and changing along the way the synaptic strengths so that the frustration of the desired patterns gradually diminishes. In the end, these patterns will be a lot less frustrated than all the other possible patterns. The memories now lie like hollows between dunes in a landscape of frustration.

As early as 1949, the psychologist Donald Hebb presented a very simple hypothesis of how synaptic strengths change in biological neural networks. He proposed that if two neurons in an activity pattern were both active, then a process would occur which would gradually increase the synaptic strength between them. Thus, the neurons' influence on each other is strengthened and, in the future, the neurons will have an increased tendency to fire simultaneously. This process would clearly serve to diminish interneuronal frustration.

Thus, in a symmetrical neural network, there are two different frustration-diminishing processes: the first is the dynamic development during which a remembered activity pattern is recalled; the second is the learning process by which a given activity pattern is stored in the network by adjustment of the synaptic strengths. Although these processes are often separated in model calculations, in principle, there is no reason why they should not take place simultaneously.

The learning process could be described as an unsymbolized translation of the required memory into synaptic strengths and, conversely, recall can be seen as the reverse translation of synaptic strengths into the activity pattern of the memory.

Most translations of information between different representations are *local*, because a small portion of the untranslated text corresponds to a very particular part of the translation. This applies, for example, between such natural languages as Danish and English, and to the telegraphic Morse Code. Conversely, in the neural network, stored information is *distributed* over all the net's synaptic strengths, and it is impossible to pinpoint any one place where a particular piece is stored.

This distributed nature of the memorized information at the same time ensures the robustness of the memory. It turns out that, even if many of the network's connections are destroyed, the minimum of frustration — the hollow in the dunes — which corresponds to a particular memory, does not move much. Even after rather extensive destruction of a neural network it is, in fact, possible to recall its memories more or less perfectly. Perhaps this robustness explains the resilience of the human brain to the violent assaults of alcohol, narcotics, medicines and physical trauma such as those received during moderate bouts of boxing. Compared with the vulnerability of a conventional computer, it takes relatively large amounts of abuse to permanently damage the brain. However, the manufacturers of paints and varnishes have discovered a trick through the use of organic solvents in their products.

A neural network can have a frustration landscape with more than one minimum and, therefore, more than one memory. Memories are laid down like the hollows in a landscape of dunes and the dynamic recall process is like a ball rolling down into the bottom. Each hollow has an *area of attraction* for balls landing nearby. To recall a particular memory, it is necessary to start the process, throw the "ball", within the area of attraction for that particular memory.

Information from several memories is coded in the same setting of the synaptic strengths. Naturally, there are definite limits to the number of memories which can be packed into a network of a certain size. Both theoretical calculations and computer simulation have shown that the limit of the binary symmetrical neural network is

14%, so that a 1000-neuron network can contain 140 randomly chosen memorized activity patterns. The network's capacity depends on the appearance of the memories and on the learning rule, in this case the Hebbian prescription. If the memories closely resemble each other they tend to interfere. With the simple version of Hebb's principle at most 14% can be memorized, whereas with a more complicated learning scheme that takes the structure of the memories into account, more can be learnt.

THE MEMORIES EVAPORATE

The transition at the 14% level is abrupt and dramatic. Below the 14% limit we have a well-functioning, associative memory, whereas beyond that point the network becomes completely unusable and is unable to recall a single item. The network, in fact, has two *phases*, as different as water and steam, themselves two different manifestations of the chemical compound H_2O.

In one phase the network can recall; in the other it cannot. Right up until the phase transition occurs the network's error percentage is less than 1.5%; immediately after, it escalates to 50%, which is, regardless of the memories' appearance, the error that is committed with random guessing about the binary neurons' activities. What happens at the phase transition is that the hollows of the dune landscape are washed out as if by a flash flood. When the flood recedes, it leaves an unintelligible washboard pattern in the sands.

Theory and computer simulation have also shown that a network in the remembering phase has some frustration minima that do not correspond to any of the learnt patterns. In the dune analogy, these hollows have not been "dug" on purpose by the training process, but the "ball" can, nevertheless, run into them. Some of these false minima appear as illogical or hallucinatory combinations of correct memories. For example, if the network has been taught to remember letters, the false minima may well produce a memory combination of A, B and C mixed together, not resembling any letter at all. As the number of learnt patterns gradually approaches the 14% limit, the amount of false recall also increases until a saturation point is reached where the memory completely breaks down.

UNLEARNING

The false minima are not as deeply engrained as true memories in the frustration landscape. In 1983 John Hopfield and his colleagues tried to get rid of them by taking the network through an additional learning process which would slightly weaken all the minima. This would correspond to emptying a bucket of sand into each hollow between the dunes, however small it may be. Naturally, this would have the greatest effect on the small, false hollows, thus eliminating their annoying influence. To find these small hollows the ball is thrown repeatedly at random and then when it stops rolling, it is followed up with the sand bucket. In the network, this "ball throwing" corresponds to initiating activity patterns at random, then tracing where the dynamics leads them. Next, carefully using Hebb's principle in reverse (with the sign opposite to that used for proper learning) all the synaptic strengths are thereby slightly modified.

This reverse learning process may reasonably be called *unlearning*. Of course, if this unlearning process is pushed too far, then the genuine memories will also be eradicated, which is, naturally, not the intention.

DREAMING

Science knows little about the function of sleep and less of dreaming. Do we sleep in order to dream or dream in order to sleep?

Sleep has several stages characterized by different types of brain activity. During *dream sleep* the brain displays surprisingly high activity. Is the purpose of dreaming merely to provide material for evening classes or does it have a more useful, biological function?

Dreaming takes place during periods of strong eye activity. The eyes move rapidly back and forth under the closed lids as if the sleeper is watching an exciting tennis match. This type of sleep is called rapid-eye-movement (REM) sleep. Since the early 1950s REM sleep activity has been recorded by means of electrodes attached to the eyelids. In conjunction with electroencephalography — EEG — it may be observed that the brain's activity corresponds to or even exceeds that of the waking state. On the other hand, the body's muscles, especially those of the head and neck, are more relaxed during REM sleep than they are during non-REM or normal sleep. REM is not unique to man, but is common to all mammals, even the blind moles who have considerable difficulty with eye movement. Birds also have REM sleep, but no traces of this phenomenon have been found, however, among fish or reptiles.

Normal adults dream for almost 2 hours in every 24, or spend roughly one twelfth of their lives dreaming. Every adult spends almost a whole month of every year in the dream state — not including day dreaming. Newborn babies dream for up to 8 hours daily and fetuses even more. Such a regularly recurring involuntary activity which demands much more of our lives than, say, sex for example, must play some vital role in the brain's function.

Scientifically, it seems natural to assume that dreaming is a spinoff effect of necessary neurophysiological processes in the central nervous system. This is confirmed by the fact that, contrary to the beliefs of the great Viennese shrink, Sigmund Freud, people normally forget most of their dreams. Only in rare cases, where the dreamer begins to wake up, will the tail end of the dream be remembered and even then with great difficulty. It seems quite difficult to oppose the natural process that wants us to forget the dreams.

In 1983 the biologists Francis Crick and Graeme Mitchison propounded a theory that dreams and dreaming were linked to various memory functions. The theory proposes that dreams are necessary to the stabilization of memory and recall through an unlearning process similar to that which Hopfield and coworkers utilized in artificial neural networks. During dreaming, the normal input channels as well as consciousness are disconnected, and the brain forms associations with vast numbers of memories by means of randomly created activity patterns. When a stable pattern, which may well be an illogical and hallucinatory mixture of real memories, is established in this way, the learning process is reversed and the memory's electrochemical strength is diminished. The dream is gradually forgotten, possibly after all or part of it has been dreamed several times.

If awakened in the middle of a dream and, thus, aware of the stable activity pattern, one experiences the unlearning process and it requires considerable effort to stop the dream from being rapidly forgotten. Dreaming should be regarded as a ripple on the surface of that whole profound biochemical process which is concerned with discarding meaningless mixtures of memories from all levels of the brain. Dreams thus result in a form of selective memory loss whereby the false, unstructured, and hallucinatory are thrown out. We dream in order to forget!

It was stated earlier that information processing generally has a great deal more to do with discarding the irrelevant than storing the relevant. The analogy used in the chapter on information processing was that of the sculptor who removes marble chip to reveal the statue in the block rather than trying to build it up from pieces of the same material. Dreams are a natural link in the brain's information processing pattern, rather as garbage collection is a link in the activity pattern of urban life.

Nightly dreams are similarly a waste product of the brain's working processes. This cerebral garbage is doubtless just as informative of what is going on in the brain, as are the city garbage cans of the patterns of modern urban life. For the archeologist, the garbage dumps of the Stone Age are vital information. Traditional Jungian and Freudian dream analyses place great weight on the details of dreams' content and the chains of events. In light of such modern theories as that of Crick and Mitchison this may well be meaningless. If dreams are no more than

Figure 14: The spiny anteater from New Guinea in the process of eating worms. This egg-laying mammal does not have REM-sleep. Compared to its body weight it appears to have a large brain which is reflected in its head size, but it does not have a comparatively high level of intelligence.

arbitrary combinations of memories, it may be a waste of time to try to ascribe profound meanings to them. If one finds a doll with jam smeared all over its face in a garbage can, this does not mean that dolls eat jam. Rather, it may just be a consequence of the sanitation department's old-fashioned policy of not demanding that garbage be sorted into organic and inorganic materials.

The Crick–Mitchison theory is based on some physiological observations. Our brains seem, for example, to be equipped with an activity generator in the brain stem which generates random-looking unstructured patterns while we dream. Decisive experiments in this area are difficult to conduct. Attempts made to assess changes in memory performance induced by preventing persons from entering dream sleep, either by means of drugs or simply by waking them, have produced no clear results. Drugs have unpredictable side effects, and waking people up leaves the problem of whether the effects derive from the lack of dreaming or from lack of sleep!

Some observations from the Animal Kingdom may lend weight to the theory. In Australia and New Guinea there is an egg-laying mammal, the spiny anteater, which displays no REM sleep at all. It also has a cerebral cortex which is abnormally large in proportion to its body weight. This anteater has been a source of irritation to neurologists for almost a century (though, no doubt, to ants for far longer) because it displays no manifestations of intelligent behavior beyond the instinctive level. It has, in other words, very limited learning ability. Perhaps the anteater's way of avoiding memory breakdown is to have an oversize neural network. Perhaps the animal can manage with a very large brain and an uneconomic memory that does not utilize unlearning to get rid of parasitic memories.

It appears that REM sleep may provide animals with a more effective brain of a smaller volume than would be necessary without the garbage-removal function of dreaming. This contention is supported by Russian investigations from 1984 which found that two dolphin species with large brains apparently did not experience rapid-eye-movement sleep, whereas that is usual for two seal species. Both dolphins and seals are clever enough to make a living in circuses, but have different brain sizes. One could speculate that the seal's need to stick its head up through small holes in the ice to breathe has forced evolution to shrink the seal's head and thereby its brain size as much as possible without affecting its intelligence level.

It is impossible to say whether this theory of dream function will stand up to the critical attention it is bound to receive, but it is an example of a scientific explanation from the borderland between neuroscience and psychology. Its point of departure is in the microscopic properties of the brain's neural network and it gives a hint of how such a macroscopic phenomenon as dreaming may be interpreted. This type of approach, crossing traditional scientific boundaries and bridging levels, is characteristic of the contemporary research surrounding artificial neural networks.

PERCEPTRONS

The ability to discriminate between categories is an indispensable ingredient of intelligence. It is scarcely surprising that this cognitive capacity is among the first which man has attempted to imitate. In the "zoo" of neural networks there is a special species which is particularly well suited for simulating the human ability to discriminate.

Every day, all of us make definitive choices between situations which are, at first glance, remarkably similar. In 1767 the composer Johann Schobert killed himself and part of his family and some friends by demanding a basket of poisonous fungi prepared at an inn on the outskirts of Paris. Distinguishing between poisonous and nonpoisonous fungi is both difficult and dangerous. It is still a recurrent theme in France that whole families die because of the patriarchal dictum "Papa knows perfectly well what edible fungi look like". Nevertheless, it is possible with careful training and practice to make very fine distinctions between categories of fungi that look almost the same.

The human classification process is an association between sensory perception and corresponding categories. This process is part of *perception*. The number of possible sensory impressions is infinitely large. No two fungi are exactly alike, but still it is necessary to be able to determine the toxicity of arbitrary specimens. The number of categories can also be very large. When we see someone's face we must be able to decide if it is known to us and, if it is, put a name to it.

Reading aloud is a perceptual process whereby sight is converted to sound. Even though at school we form the impression that it is a simple translation of letters into elementary sounds, the process is, in fact, much more subtle. Pronunciation does not depend on a letter alone, but on its context: the surrounding letters, perhaps the place of the word in the sentence, and even the sentence's meaning.

In the sentence "I read the Bible" it is not possible to guess from the sentence alone, how "read" should be pronounced. In the present tense you could be going steady with God; in the past tense you may have turned your back on him many years ago. From the written form of this statement the parish clerk cannot determine whether or not there is an obvious opportunity to squeeze you for a contribution.

The principal characteristic of mental classification is the establishment of a connection between two patterns: a sensory impression and a category, an input and an output, a question and an answer. In the human brain both input and output are represented by neural firing patterns and the connection between them is made in the brain's neural network.

Artificial neural networks used for classification often have the same general structure. A popular architecture called a *perceptron* is like a sandwich, with a bottom, a filling and a top. The perceptron's upper layer consists of artificial input neurons receiving their input from the user, the lower layer is composed of output neurons which produce the answer. Perceptrons communicate externally only via these two layers of neurons. The connection between them is established through a *hidden* network of artificial neurons, the filling in the sandwich, and the principal distinctions between perceptron types are structural variations of the hidden part. Generally, the term perceptron is reserved for systems in which communication between input, hidden net and output is one-way; there is no *feedback*. This is the radical distinction between perceptrons and symmetrical binary networks which derive their special skills from feedback.

That artificial neural networks could be used to solve classification problems was first shown theoretically in 1943 by Warren McCulloch, a neuroscientist,* and Walter Pitts, a mathematician. They introduced the idea of formal neurons which correspond to the binary "all-or-none" neurons discussed in the previous chapter. The binary neurons also bore a certain resemblance to the picture one at that time had about the functioning of real, biological neurons. The important point was that McCulloch and Pitts showed that, in principle, their formal neuron network was capable of solving any problem of logical character. However, this was only in principle and no method was indicated as to how a network can be built to solve a particular problem in the most economical way. In recent years, research in neural networks has diminished the problem, but still without finding a definitive solution.

---

* The impressive professional activities of Warren McCulloch (1898–1969) cannot be described with a single designation of occupation. At various points of his life he worked in mathematics, philosophy, neurology, neurophysiology, psychology, and psychiatry. He was also an accomplished poet.

ONE-LAYER PERCEPTRONS

The simplest perceptron has no neurons in the hidden network, only synaptic junctions between input and output. Usually, every input neuron influences every output neuron. It is the strengths of this *single* layer of connections which alone determine what the perceptron is able to do. The input neurons contribute little to the perceptron's skill because they only pass on the activity generated in them by the surroundings, whereas the output neurons perform a real calculation on the basis of the collective influence exerted by the input neurons.

Figure 15 shows the three simplest one-layer perceptrons. They play an important role in understanding the much more complex structures that can be built by extending the network both in "breadth", with more neurons in each layer, and in "height", with increased numbers of hidden neuron layers.

The simplest perceptron (Figure 15a) has only one input neuron and one output neuron. It can do very little, but is nevertheless capable of simulating each of the 1-bit computers referred to in the chapter on information processing. The simulation is achieved by varying the synaptic strength and the threshold of the output neuron; as the input performs no calculation its threshold is irrelevant. The only 1-bit computer which performs any nontrivial function is the one which exchanges 0 and 1. This is called NOT and is realized by, for example, selecting synaptic strength –1 and threshold –0.5.

This can be understood in the following way: when the activity of the input neuron has a value 0, the "collective

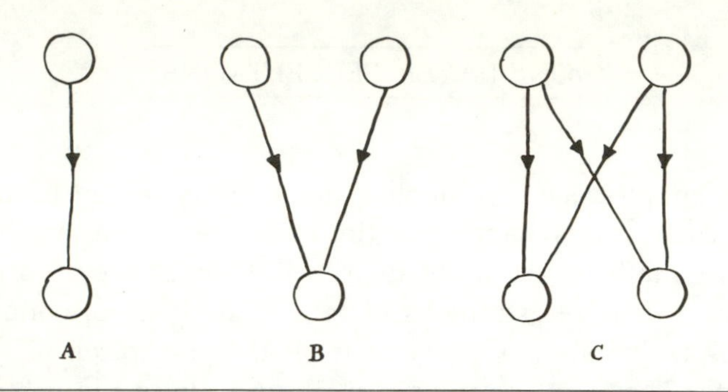

Figure 15: a) The simplest possible perceptron has one input and one output neuron; b) The second simplest has two inputs and one output; c) A perceptron with two input and two output neurons.

pressure" on the output neuron will also be 0, and because this is greater than the threshold -0.5, the output neuron's activity will be 1. Conversely, with an input neuron activity of 1, the collective influence on the output neuron will be -1, which is less than the threshold value of -0.5, and consequently induces an output neuron activity of 0. Thus, the perceptron's function is: $1 \rightarrow 0$ and $0 \rightarrow 1$, or NOT. This example demonstrates that one-layer perceptrons are capable of simple classification corresponding to logical functions.

There is more power in the logical functions that the second perceptron (Figure 15b) can perform. It has two input neurons and one output neuron. The two inputs are evaluated and the output is produced as the result of this evaluation. As there are 4 possible input and 2 possible output configurations this means that there are $2^4 = 16$ possible classifications or logical functions of this type.

Of these 16, it turns out that only 14 can be realized with this simple one-layer perceptron. Unfortunately, two very important logical functions are missing: XOR and NOT

XOR	NOT XOR
00 → 0	00 → 1
01 → 1	01 → 0
10 → 1	10 → 0
11 → 0	11 → 1

Figure 16: XOR and NOT XOR, the two logical functions which cannot be realized in a one-layer perceptron.

XOR (Figure 16). These functions distinguish between inputs where the two neurons have the same activities and those where they are different. The logical function of XOR is to choose one or the other, but not both.

All one-layer perceptrons of the type in Figure 15b can put the 4 different input possibilities into exactly 2 output categories. If the 4 inputs represent 4 varieties of fungus (00 = king bolitus, 01 = destroying angel, 10 = fly agaric, 11 = field mushroom), the two outputs — 1 and 0 — could represent toxic or nontoxic. As both the first and the last mushrooms are nontoxic, while the middle two are toxic, the XOR function will discriminate correctly between them. This discrimination is a textbook example of non-linear classification of which the one-layer perceptron is incapable. It is necessary to make the network a little more complex to separate the fungi correctly.

Input, either to the brain or an artificial network always has a *representation* in the form of some neural activity pattern. The four types of fungi above have been given an explicit representation in which those which closely resemble each other also have closely resembling activity patterns. Thus, the fly agaric and field mushroom patterns diverge by only one bit because they look alike. Their toxicities, however, are very different.

The cognitive difficulty here is to make a sharp distinction between fungi which closely resemble each other but have widely different toxicities. This type of cognitive problem crops up every day in one way or another. How to choose between the good and the bad doctors when both wear white coats and write equally illegibly? How do you tell that an apartment has only been given a quick face-lift and is not worth the asking price? How do you see through the paint job on a used car?

In the brain, Nature has chosen the representation for us through the construction of our sensory apparatus. Spectacles and hearing aids can preprocess information, but we cannot do much about the way our retinas catch light. To make subtle distinctions in our experiences, it is necessary to use a large measure of XOR-type discrimination.

Comic books and old gangster movies also preprocess information for us: the good guys are handsome, the bad guys are ugly, but in reality and in some recent movies it is a lot more difficult to tell the white hats from the black.

In 1969, the computer scientists Marvin Minsky and Seymour Papert, in their book *Perceptrons*, demonstrated that no one-layer perceptron could solve XOR problems. This mainly theoretical but also political book had a great impact on the emerging neural network community. They daringly extrapolated their analysis of simple one-layer perceptrons to imply that research in multilayer perceptrons would be sterile, and their negative attitude did a great deal to cool the interest in perceptrons and neural networks in the following years. In the late 60s the victory went to the serial computer worldwide and artificial intelligence was reduced to the level of advanced programming in special logical languages. As a research area massively parallel computation slid into the background and was only pursued by a few "fanatics". It is thought-provoking that the pessimism of two influential scientists could affect a whole generation of research in computer construction and use.

One can understand why the second perceptron (Figure 15b) cannot realize XOR, because the output neuron adds the influence from the two input neurons before it decides which category to choose. If we suppose that the output neuron threshold is selected positive then the 00 input cannot activate the output which is as it should be. When the influence from one of the input neurons in the XOR case — 01 and 10 — is able to activate the output neuron, then the sum of their influences will also do so in the 11 case. Thus, it is only possible to obtain the correct answer from a one-layer perceptron in 3 of the 4 instances in the XOR table, a situation which could prove fatal if you are relying on the perceptron to choose your mushroom.

There is nothing in the way for the implementation of other logical functions such as OR or AND. Manipulating the two synaptic strengths and the threshold one can, as already stated, make the perceptron in Figure 15b realize 14 of the 16 possible logical functions of two variables.

HIDDEN LAYERS

The complexity of the one-layer perceptron is not sufficient to handle XOR categorization problems. The logical depth of such problems is too great for a "flat" perceptron to cope with. If, however, an appropriate number of hidden neurons is interposed between the input and the output, then all logical functions can be implemented.

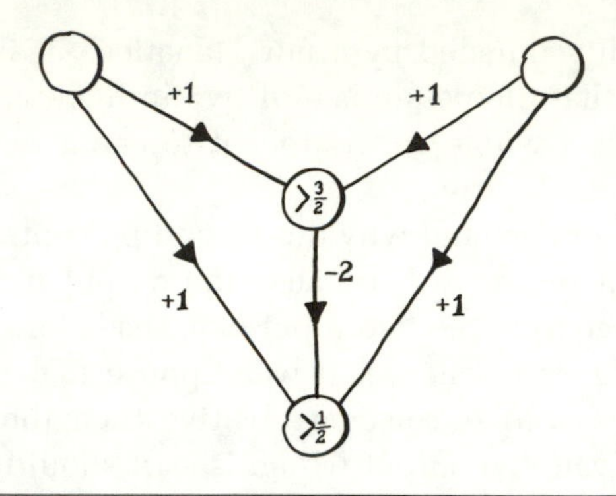

Figure 17: The simplest perceptron with hidden neurons realizes every logical function of two variables. The strengths and thresholds shown in this figure correspond to the function XOR.

Figure 17 shows a perceptron with two layers of synaptic connections. This network can cope with XOR with the indicated synaptic strengths. Note that the hidden neuron possesses a threshold value which only allows it to be activated by input 11. When active it has a strong negative influence on the output neuron in the lowest level. When the hidden neuron fires, this negative influence cancels the positive effect from the two active input neurons and thereby induces the same total effect on the output neuron in both case 11 and case 00.

The hidden neuron is only activated by input pattern 11, and it can therefore be said that it looks for a particular *feature* in the input pattern to which the perceptron is subject, namely, that both input characteristics are present.

The hidden neurons in multilayer perceptrons can be arranged in innumerable ways to suit the nature of the problem. Currently, there is no standardized mode by which the optimal perceptron can be tailored to a given

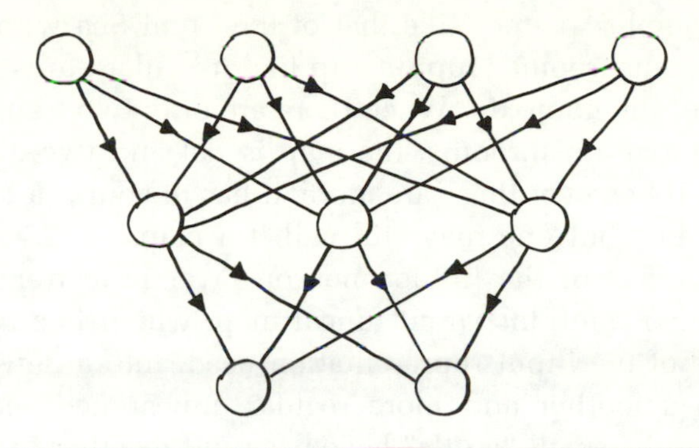

Figure 18: A feedforward perceptron with two layers of synaptic connections. In this particular perceptron there are no mutual connections between the hidden neurons and no direct connections between input and output. Otherwise the perceptron is fully connected between the layers.

problem. A large part of contemporary research is directed towards finding rules of thumb for network architecture as well as a proper theory.

Figure 18 shows one of the most frequently used constructs which has only a single layer of neurons concealed between input and output. In this design there are no mutual connections between the hidden neurons nor any direct connection between the input and output levels. Since information only goes in one direction from input to output such networks are called *feedforward*.

If a perceptron were really going to be used to discriminate between fungi then it would need a great number of input neurons capable of giving a more accurate description of actual fungal characteristics. In the small network in Figure 17 only two features could be taken into account. If one was to make a "map" of all the possible fungi which the perceptron might encounter it would have a very complex pattern like that of the Coral Sea with poisonous mushrooms cropping up like irregular atolls in an ocean of the nontoxic. When there are only two traits that characterize mushrooms, the map is flat and two-dimensional. If however there are more, it has of course a higher dimension, but one may still call it a map.

The effect of the hidden neuron layer is to transform this rather nontransparent fungal map, which is a consequence of the input representation used, into a new map with a smoother and more regular appearance because many of the small "atolls" have been put together to make larger land masses. The new map need bear no relation in size to the old, but is merely another representation which is easier to find one's way in when correct decisions have to be taken about toxicity.

If the perceptron is to carry out pattern association with great logical depth, also called *strongly nonlinear* association, several layers of hidden neurons may be advantageous. Strong nonlinearity is characterized by questions that, although closely related, frequently have answers that differ a lot. Insurance policies are often highly nonlinear so that although "open theft" may be covered, having your wallet stolen by a pickpocket may not.

In the feedforward perceptron, information streams from layer to layer, being "decanted" and filtered until, at last, a map is produced in which every answer category is neatly separated from all the others.

BACK-PROPAGATION OF ERRORS

Very small perceptrons like XOR in Figure 17 contain so few adjustable parameters that useful values can be found by hand. Perceptrons for use in more realistic calculations always contain a very large number of parameters: synaptic strengths and thresholds.

With all strengths and thresholds set at 0 the network's output will be trivial because all inputs are put into the same class characterized by all output neurons having activity 0. Another uninteresting limit occurs when all the strengths and thresholds are set random which produces a classification which is useless. Synaptic strengths and thresholds must be related to the problem to be solved.

Apart from very special cases, most pattern associations can only be described through the use of examples of the correct classification. Attempts to formulate rules easily lead to a situation of making endless exceptions to exceptions to exceptions. This can occur, for example, if one attempts to formulate a complete set of rules for division of English words at the end of printed lines, like it is done in this book.

To get a network to correctly classify a set of examples is a difficult exercise in itself, the complexity of which naturally depends on the degree of nonlinearity inherent in the problem. If there are many closely related examples which have to be put into widely different categories then the task will obviously be much more difficult than if they can be easily arranged in neighboring categories. In the word division case, the words `demonstration` and `demonstrative` differ only in the two last letters, but have to be hyphenated quite differently: `dem-on-stra-tion` and `de-mon-stra-tive`. The reason is, of course, that the pronunciation is quite different, but it illustrates the presence of nonlinearity in the problem.

Given a known set of examples that correctly connect input and output, object and category, these can be used to train the network in much the same way that people learn to divide words. The method is to adjust the network's many parameters in light of the mistakes made in the training examples. If the network gives the wrong answer, one, so to speak, sends the mistake back through the network. This *back-propagation** adjusts the values of all the strengths and thresholds so that next time the network is shown the same example it will be closer to producing the right answer than before. If the network is sufficiently large, after it has encountered a suitable number of training examples it will be able to classify them correctly.

GRADED NEURONS

There is an ostensibly, purely technical detail which is a necessary prerequisite for the functioning of this supervised learning process which works by back-propagation of errors. During training the classification errors made by the network are calculated. At the start, before the network has learnt, these errors are very large. Even though we know the amount of error the network makes, it is not possible with current methods to carry out a large adjust-

* This is a terrible construction, but is the common one used by workers in the field. To *propagate* is to move something forward.

ment in the network's parameters so that the error can be eliminated in one go.

The error can only be used as a bearing to indicate in which direction the correct parameters appear to lie. Under the influence of all the different associations which we try to imprint on the network, the direction of the bearing will change many times in the process. Some associations strengthen each other because they have many common features; others cause disturbance because they discriminate between fine details in the input.

One is thus forced to alter the network's parameters in very small steps corresponding to a fraction of the error committed. It is impossible to alter the activities of binary neurons that only accept values of 0 or 1 by less than the full difference between these activities, which amounts to either +1 or –1. Small changes of the network's output in the right direction are just not possible.

The binary neurons that only have activity at 0 or 1 are, therefore, no good in this respect. Instead, neurons are introduced that can have any activity between 0 and 1, for example 0.4711. These *graded* neurons have a more nuanced response to "collective pressure" than their binary cousins. The variable response shown in Figure 19 has a characteristic S-form, which in reality has much in common with the response of binary neurons. The curve is also very nonlinear. It remains close to 0 while the influence is under the threshold, but rises sharply to 1 once the threshold is crossed. There is only a very small interval of influence where the answer from these neurons is "Perhaps". It is due to the nonlinearity that they have retained the ability to make clear decisions. The nonlinearity is still confined to a very narrow interval of influence, which makes it possible to have nonlinearity in the

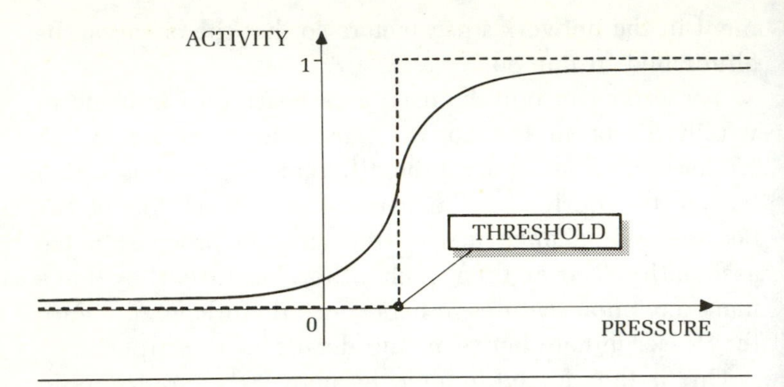

Figure 19: Input-output relation for a neuron with a continuous interval of activity levels. This S-shaped curve is an expression of the neuron's nonlinearity. The corresponding curve for a binary neuron is shown as the rectangular broken line. Note that the activity is limited downwards by 0 and upwards by 1.

perceptron just where it is needed.

With this graded response, the artificial neuron approaches more closely the function of the biological. No biological response can be so categorical as that of the binary neurons, and S-shaped curve responses have in fact been observed in many kinds of natural neurons.

A neuron sits, like a spider at the heart of its web, watching the activity of other neurons. It combines these activities with the synapses' strengths and works out its own activity from the combination. The supervised learning process sets up the strengths of the synapses so that each neuron is capable of detecting particular features or trends in the activity of other neurons upstream in the network.

A neuron's activity may be seen as an expression of its belief in that feature it is set up to detect. The binary neurons can only establish whether a particular feature is present or not, whereas the graded neuron can express itself much more subtly. The graded neuron's activity may be viewed as the probability of the presence of a particular structural feature in the input it sees.

It is the stepwise learning process which forces us to use neurons with more "natural" characteristics, but the change has, in fact, much deeper implications. This change establishes a definitive separation between the digital symbol-manipulating computer and the analog neural network. In the analog computer the continuous degrees of freedom of a physical system are used directly in calculations instead of proceeding through an intervening digital level. In an analog computer a number may be represented by the size of an electric current instead of a finite set of 0's and 1's. Whereas the digital computer always works with finite precision, the analog computer has, in principle, precision as high as desired. In practice, however, the precision is limited by the problems of measurement of physical quantities and the inevitable noise.

While the binary neural network is a type of hybrid between digital and analog computers because the activities are binary and the strengths and thresholds continuous, the graded neural networks are purely analog. The features or concepts which graded neurons represent would be very difficult to formulate in terms of symbols like the words of the English language. It is precisely at this point that the strength of the neural network is demonstrated. Neural networks are capable of dealing with concepts and combinations of concepts which are quite alien to a symbolic analysis of a categorization problem. In the final chapter we shall take another look at the philosophical consequences of this difference, which possibly has considerable significance for understanding the nature of intelligence.

GENERALIZATION

People have an extravagant capacity for *generalization*, of which this statement is, of course, a perfect example. However, without this ability all knowledge would have to be learnt as separate items or unique cases. We experience this with irregular verbs where their lack of system makes it difficult to determine their conjugation. Children learning to talk frequently use regular generalizations of conjugations they do not know. They may say "teached" instead of "taught" and "sweeped" instead of "swept".

When people are taught by means of examples, the intention is that they will begin to work out the underlying systematics for themselves. A teacher will have great difficulties explaining the detailed systematics behind word division in English. Generally, in the Western cultures we have a tradition for attempting to explain to each other the underlying systematics rather than using large numbers of examples. In Japanese education it is on the contrary usual to leave the generalization to the individual. It is very difficult for a European to learn the Japanese board game *Go* from books, because the seemingly endless iteration of one example after another is unbearable without any explanation of principle. Later, however, one comes to understand the theory from the examples.

The idea behind neural networks, and perceptrons in particular, is that by training they teach themselves to generalize from the examples they are given. When a trained perceptron is presented with an input that was not included among the training examples, it will nevertheless make a definite offer of a category. The perceptron has no means of going on strike and must come up with an answer.

The question is whether that answer is worth anything. The perceptron's answer results from the gradual adjustment of parameters which has occurred during training. Mathematically, the perceptron has interpolated between the examples it has seen. The more examples it has been given, the less is the freedom of interpolation. If the perceptron's parameters of adjustment can accommodate the principle behind the examples, once it has been exposed to a suitable number it will start to generalize sensibly.

The perceptron adapts itself to the principle behind the examples by adjusting its synaptic strengths and firing thresholds. The construction of perceptrons is a balancing act between several different factors. In the main, one never knows the complexity of the knowledge one is attempting to impart to the perceptron. Thus, neither does one ever know exactly what architecture the perceptron should have to enable it to generalize perfectly.

Neither is it always crystal clear what the correct generalization is. One of the so-called intelligence tests used by the military on prospective recruits involves the continuation of number sequences. Most people can cope with 1, 2, 3, . . . whereas 7, 9, 13, . . . however, proves generally rather harder. The problem is that each sequence has in fact many sensible continuations, the correct generalization, therefore, being somewhat a matter of taste. A number theorist might well score very badly in such tests because his ability to generalize is completely different from the military standards.

The first of the examples is, no doubt, intended to weed out the worst applicants. The desired sequence extension would presumably be 1, 2, 3, 4, 5, . . . but one might equally well use the Fibonacci sequence 1, 2, 3, 5, 8, 13, 21, . . . where each number is the sum of the two previous ones. In the second example the second sequence might be continued in the form 7, 9, 13, 19, 27, . . . by the addition of two to the difference between each step, but it could equally well be 7, 9, 13, 21, 37, . . . where the difference is doubled between each step. It is not clear what is required in order to bring home a good service record!

A perceptron is too small if it cannot be trained to repeat all the training examples correctly, even when repeated frequently enough. Even though it is too small, it may hit on a generalization which contains at least a certain element of sense, but where not all the training examples can necessarily be learnt in detail.

When a network is well trained, it can ignore minor inconsistencies and mistakes in the training examples and still understand the correct systematics. If there is direct conflict between training examples, so that the same input corresponds to different outputs, the network will become confused in much the same way as are children of divorced parents who are unable to agree about anything.

A problem can also arise if the perceptron is too large. It will take as law the small mistakes and discrepancies, which are inevitable in any set of training examples drawn from real life, and the network will tend to reproduce them, instead of the systematics. The network then becomes, in a sense, overspecialized and, like some experts, it will place too much importance on the small details of its particular field of competence. The network behaves like *les nouveaux riches* who have suddenly come into a lot of money and go a bit over the top in everything.

Today, the training of neural networks is more an art than an analytically founded science. It is known that supervised training of networks is in the category of hard optimization problems like the *Traveling Salesman Problem*. This is the problem of finding the shortest route connecting a number of cities each of which has to be visited once. When the number of cities is more than about 20 and the cities are scattered randomly, it is usually impossible to establish the optimal route manually within a reasonable time. The problem is hard because the number of required logical steps increases all out of proportion with the increase in the number of cities.

Training a network is a slow process which on real-world data imposes a limit on just how much the net can learn from the examples. Training a network to make diagnoses on the basis of symptoms could very well involve as many logical steps as training a doctor. However, because of the speed of electronics compared with biological circuitry, it may be faster for the electronic networks of the future to learn to make diagnoses than for the doctors.

On the other hand, the use of a trained perceptron is quick, because it does not require lengthy adjustment of parameters, but only a calculation of the neurons' activities. The asymmetry between the learning and use of what has been learnt is familiar from ordinary education. A tiger act in circus is very brief compared with the length of time it has taken to train the tigers — which also includes the time taken in teaching the tigers not to sink their teeth into the trainer.

APPLICATIONS OF PERCEPTRONS

The perceptron is undoubtedly the neural network currently most applied in practical situations. Pattern recognition of one form or another is involved in a wide cross-section of realistic problems. The following sections will present two examples, word division in English and a network that learns to read aloud from a written text.

Of the almost innumerable applications in many spheres, we shall name a few of the most recent: identification of handwritten numerals in zip codes, prediction of future values of currencies, detecting explosives in checked luggage from gamma-ray spectra, finding structure in the time evolution of Bach cello concertos, determination of the spatial arrangement of surfaces, the coordination of robot sight and motor control, mortgage risk evaluation, submarine identification, prediction of the three-dimensional structure of proteins, speech recognition, veterinary quality control of meat, winning strategies for backgammon, and food collection in artificial animals.

WORD DIVISION

When writing a memo or typesetting a book, one eventually comes to an end-of-line, because the paper is not infinitely wide. At this point there may be space enough on the remainder of the line to write part of a word, but not enough to complete it. It is bad style to let the word stick out into the right-hand margin, and it is also considered ugly to space out the other words on the line in order to consume what is left. So if the page is to appear professional, one is from time to time forced to divide words at end-of-line. This is usually done by means of a hyphen signaling that the word at the end-of-line is not the whole story.

It has been mentioned before that hyphenation requires intelligence, and that this problem cannot easily be solved by formulating a compact set of symbolic rules. The native writer knows how to hyphenate most words in the same intuitive way as he knows how to ride a bicycle. Even then doubtful situations often occur. Do you hyphenate the word `handling` as `hand-ling` or `han-dling`? Do you write `accord-ance` or `accor-dance`, `hom-icide` or `ho-micide`? Is it all right to hyphenate `ask-ed`, `al-so` and `o-pen`?

If you demand somebody to pin down the rules, it will quickly become clear that there are exceptions beyond exceptions. In English, words are hyphenated according to their pronunciation, most people and even linguists would claim, but this only shifts the problem to formulating the rules of English pronunciation — and that everybody agrees is a hard problem. In some texts on hyphenation in English, the advice is to consult a dictionary rather than "to trust pronunciation or try to memorize the principles".

The problem of hyphenation has a character which makes it natural to try to use a perceptron, and in 1988 this was done by the authors of this book. The classification task which the perceptron is required to perform, is to decide whether or not it is correct to hyphenate at a point between two consecutive letters in a word. This decision shall be made from the context of letters surrounding the point in question. This context is chosen to be rather limited, consisting typically of four letters on each side of the possible hyphenation point. Thus the perceptron views the word through a *window* eight letters wide.

This brings us immediately face to face with the two central problems in applying neural networks to real-life situations: the problem of *representation* and the problem of *architecture*. How should the letters be represented in terms of neural activity patterns, and what should the hidden part of the network look like?

In English there are 26 different letters and the network should be able to discriminate between them easily. This requires that there be a significant difference between the activity patterns which represent the letters in the input layer of the network. It is therefore reasonable to represent every letter by a block of 26 neurons, of which only a single one fires, whereas the others are inactive. In this way each of the 26 letters receives its own private firing pattern.

One might out of stinginess be tempted to use the continuous activity interval of the graded neurons to subdivide it into 26 smaller subintervals, each representing a single letter. But this approach would fail miserably, because letters that are close to each other in the alphabet would thereby exert almost the same influence on the hidden part of the network, and the hyphenation problem would thus be presented to the network in a form which is much more nonlinear than necessary. From a linguistic point of view there is no close relation between the letters `a` and `b`, even though

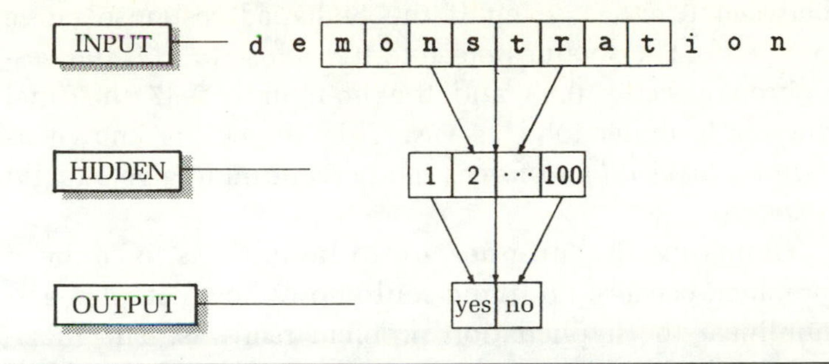

Figure 20: The hyphenation perceptron. Each input block contains 26 neurons and represents a single letter. At a given time the perceptron sees 8 letters in the window, 4 letters on each side of the possible hyphenation point. The hidden layer contains 100 neurons and the output layer only 2, representing yes and no. The network is completely connected in the forward direction between the layers. The word is "clicked" through the input window letter by letter and in this way decisions are collected about all possible hyphenation points in the word.

they are neighbors in the alphabet. With respect to hyphenation, it is better for all letters to have the same mutual distance and this is realized with the first representation.

If one chooses to have 8 letters in the input window, the input layer of the perceptron will contain 8 blocks each with 26 neurons, altogether 208 neurons. In Figure 20 the architecture of the perceptron is sketched. The perceptron has only two output neurons, each representing two categories of answers: *yes* if the word can be divided, and *no* if it cannot. The answer given by the perceptron is decided by that output neuron, which has the highest activity. As mentioned earlier the graded neurons have continuous activities that can take all values between 0 and 1, even if the S-shaped response curve preferably takes values close to 0 or close to 1. If the *yes*-neuron answers 0.63 and the *no*-neuron 0.37 the final answer is understood as *yes*. This strategy is known as *Winner takes all* and forces the perceptron to give straight answers.

Choosing the number of hidden units is a hard problem, because no living soul knows how complex and nonlinear the hyphenation problem really is. The fuzziness of the rules bears witness to the fact that the problem has a fair amount of nonlinearity, but how big it is can only be decided by experiment. In English it seems as if at least 100 hidden neurons are necessary and this implies that the perceptron contains $100 \times 208 + 2 \times 100 = 21\,000$ synaptic strengths and $100 + 2 = 102$ thresholds. All in all the network thus contains 21 102 adjustable parameters. The structural complexity of this perceptron corresponds in a certain sense to that of a machine with 21 102 independently adjustable knobs. The perceptron is analogous to the huge mixing board used for adjusting sound at a rock concert. The "mixing board" of the perceptron must be sufficiently large to be able to accommodate the complexity of the hyphenation problem.

In some languages hyphenation is learnt as a spinoff from the process of spelling and pronunciation, because the children are taught to speak each syllable along with its spelling. In this way the boundaries between syllables are identified and this makes hyphenation a relatively easy matter. The rules are still missing, but you *know* where the boundaries are. In English this method is not used widely and it is a bit of a mystery how English-speaking people learn to hyphenate. The answer may be that many simply never learn it!

In order to train the perceptron it must be presented with reams of correctly hyphenated words. Being native Danes the authors of this book have no reliable built-in ability to hyphenate English words. So we used 15 699 randomly chosen words from a British dictionary (Collins Gem: *Word Speller and Divider*, first published 1973). In the United States hyphenation differs slightly from the proper British way, because in Britain hyphenation is determined more by word origin than by pronunciation. An Englishman might break up `accordance` into `ac-cord-ance` whereas an American might prefer `ac-cor-dance`.

This database consisting of 15 699 words and their hyphenation was used for training the perceptron. Each word was presented to the perceptron and pulled through the window letter by letter. The words contained 135 902 letters and gave rise to 120 203 different training examples, each word fragment consisting of up to 8 letters. Due to the complicated set of "fuzzy rules" hidden in the database with exceptions scattered all over, it was necessary to present the same examples many times. The hyphenation perceptron was trained by exposing it to all the examples 300 times, which makes around 36 million individual presentations.

At the beginning the network was assigned random synaptic strengths and thresholds, and its hyphenation was accordingly completely random and senseless. At each presentation of a word fragment the error was determined and used to turn the "knobs" in such a way that a subsequent presentation of the same fragment would yield an answer closer to the desired one. Eventually, after 36 million presentations of fragments, the perceptron could make 99.8% of the decisions correctly.

Without hidden neurons a one-layer perceptron could only be trained to a level of 83%. In 17% of all cases the one-layer perceptron is unable to make the right decision: it hyphenates where it should not or it forgets to hyphenate where it should. The remaining 17% of the decisions can only be made with hidden neurons and must therefore contain XOR-like nonlinear structure.

It is not much fun to teach a perceptron what one already knows, unless it can become an expert hyphenator, even with words it has never seen before. How well does it *generalize*? In order to test the perceptron it is necessary to possess another database of correctly hyphenated words that are not among the training set. It is difficult to make a balanced choice of new words, so one of the tests we have made is to use the manuscript of this very book, which contains 38 471 words. There are 5022 different words and 2110 of these are not found in the training set. These new words were hyphenated by hand using a dictionary. The trained network was then asked to hyphenate the new words and the errors were counted. Altogether the perceptron made 94.8% of the decisions correctly.

Another test to which we have subjected the network is to let it hyphenate artificial words that have the same letter frequencies and almost the same interletter pair correlations as real English words. Such *pseudo-English* words may be generated in a computer using random numbers. The words have a strange Shakespearean sound, and they are clearly related to English. Most English speakers have no trouble dividing these words, and our test shows that the network is also quite clever at this task. This test is qualitative, rather than quantitative, but shows that the network has caught the flavor of English hyphenation without having resorted to a set of rules in

prodarous	prod-ar-ous
urulonital	uru-lon-i-tal
romeraty	rom-er-a-ty
macty	mac-ty
dace	dace
radause	ra-da-use
atele	atele
depan	de-pan
cedecissivitore	ce-de-cis-siv-i-tore
phribophale	phri-bo-phale
ricoxtate	ri-cox-tate
dicaretin	di-care-tin
photore	pho-tore
diesy	di-e-sy
divomy	div-o-my
sinbamplicat	sin-bam-p-li-cat
pinis	pi-nis
stiouris	sti-ou-ris
ceasme	ceasme
vonaginfinile	von-a-gin-fi-nile
disus	dis-us
cecyx	ce-cyx
urty	ur-ty
montin	mon-tin
pheder	phed-er
phars	phars
arsy	ar-sy
paroshot	paro-shot
vongamberen	von-gam-ber-en
uliase	uli-ase
cecatiore	ce-ca-tiore
conch	conch
omatanote	om-at-a-note
analidetic	an-a-li-det-ic
arbong	ar-bong
ulove	ulove
capingralice	cap-in-gralice
missy	mis-sy
camatiale	ca-ma-ti-ale
cabefat	cabe-fat
redizogusky	red-izo-gus-ky
mee	mee
panampt	pan-ampt
umparifin	um-par-i-fin
raiacot	rai-a-cot
phatejune	pha-te-june
susfull	sus-full
uspy	us-py
mutic	mu-tic

Figure 21: Examples of pseudo-English words having the same letter frequencies and pair correlations as real English. The words have been hyphenated by the trained network and apart from a few cases humans agree with the network's hyphenation.

symbolic form. The rules have been inferred by the network itself during the presentation of all the examples and is entirely represented by the positions of the knobs, all the 21 102 synaptic strengths and thresholds. In Figure 21 a sample of pseudo-English words is shown together with the network's hyphenation of them.

It may seem strange that one can make 102 small and simple-minded computers work together to produce a single decision about a hyphenation point in a word. But that is in fact what the 100 hidden units and the 2 output units collaborate to do. The neural network architecture and the learning algorithm have solved this nontrivial management problem. Viewing the network as a small company, the 208 input neurons sit at the front desk, open the mail and pass it on without modification to the next level in the bureaucracy. The 100 hidden units act as a group of clerks, each looking for specific features in the incoming mail. Each clerk produces a memo summing up his or her excitement about the incoming message. The incoming word fragment has now been transformed into 100 "memos", each containing a single number. The rest of the information in the input has been discarded at this point. Finally two decision makers, the output units, scan all these memos and decide whether the result should be *yes* or *no*. At this point the information in the original word fragment has been boiled down to a single bit.

Information processing consists in throwing away information rather than creating it. In this example information is thrown away by 100 different agents at the same time. Each hidden unit can produce its memo without knowing what the other "clerks" will write in their memos. Because of this parallelity among the "clerks", neural networks are in principle able to process information at very high speeds. The usual von Neumann bottleneck has disappeared. If the network is set up in an ordinary serial computer, the hidden units have to be lined up in a queue and handled, one at a time. This will of course slow down the "memo-creating" process by a factor of 100 for the network discussed above. But in a machine with a large number of individual processors that may be set up to communicate with each other as in the hyphenation network, the computation may be completed 100 times faster than on a single processor.

In the human brain billions of neurons are constantly sending "memos" to each other. The network of the brain is not of the feedforward type, the distinction between "clerks" and "decision makers" is not clear at all. The brain contains innumerable feedback loops which give *time* a much more prominent position than in the naive model we have used for the hyphenation problem. In a perceptron calculations take well-defined numbers of steps, whereas in the brain they may take a lifetime.

NETTALK

When a text is read aloud a sequence of letters has to be transformed into sound. The simplest design for a reading-aloud algorithm would be for each letter to correspond to an individual sound, but this method would come to a rapid end in meaningless babble.

Pronunciation of letters is strongly dependent on their context: the `p` is silent in `psychology`, but sounded in `pickaxe`; the `g` is silent in `neighbor`, but sounded in `big`. Even if not silent, letters may be pronounced differently in different combinations: g is hard in giggle, but has a soft j-sound in gelatine; c is soft, like s, in cerebral, but hard, like k, in cat.

There are also a number of English words which are pronounced the same, but spelt differently with different meanings: lock, loch, lough are all pronounced alike, but have different meanings; similarly with genes and jeans. The meaning is determined from the context. Similarly, there are words which contain syllables which are spelt identically, but pronounced differently according to prefix and context: cough, bough, lough are all pronounced differently, as are bow in "bow and arrow", and bow in "bow to the Queen".

These examples show that reading aloud is a nonlinear process so difficult that it usually requires professional expertise for public performance. However expert, such professional actors and readers never read "straight from the page" if they can help it, but always prepare their texts thoroughly in advance. On television, despite the technical help of the autocue or teleprompt, we frequently see the newscaster choke on a word or a line, particularly if suddenly handed new material.

In 1986, the physicist Terrence Sejnowski and his student Charles Rosenberg aroused great interest with a neural network of the perceptron type which displayed considerable diligence in "reading" aloud, and which they named NETtalk.

Sejnowski and Rosenberg formulated the reading-aloud problem as a translation between two texts. One was a preprocessed version of the original written text, stored in the computer, the other was the same text transformed into a "sound writing" or phonetic transcription. The "letters" of this transcription are called *phonemes* and are characterized by a number of features such as stress, lip and tongue positions, and aspiration.

The phonemes may later be sent through a speech generator so that it is possible to hear what the perceptron is actually saying. Such speech generators are readily available commercially. It is not this last stage which presents difficulties in reading aloud, the hard part is the transformation of the text into phonemes.

Like in the hyphenation problem it is necessary to have a window into the text to be read. This window determines just how much of the surrounding text should affect the pronunciation of the central letter in the window. NETtalk's window is 7 letters wide. English has an alphabet of 26 letters and there are, in addition, three punctuation marks which affect pronunciation. The letters and punctuation marks were each represented by blocks of 29 graded neurons of which only one was active at any one time. The input was therefore composed of $29 \times 7 = 203$ neurons.

The soundscript which Sejnowski and Rosenberg utilized contained 60 phonemes which were combinations of 26 articulatory features. The output layer therefore consisted of 26 neurons each representing a particular feature of the sound produced. The activities of the output neurons formed patterns, which did not have to conform to any one of the 60 phonemes, but could be an unpronounceable, possibly conflicting mixture of the 26 sound characteristics. Which phoneme the network should select was determined by calculating the distance between the actual output and each of the 60 predefined phonemic patterns. In this way, the phoneme nearest the output gained the right to be spoken. This *Winner takes all* strategy was used to ensure the production of unambiguous answers in English.

The network architecture was of the standard feedforward type. The hidden network contained 80 neurons fully connected to both input and output layers, and contained $203 \times 80 + 80 \times 26 = 18\,320$ synaptic strengths and $80 + 26 = 106$ firing thresholds, a total of 18 426 adjustable parameters.

As a training example, Sejnowski and Rosenberg used, among other texts, a phonetic transcription of a story read by a child in first grade. The first 1024 words of its story were processed through the network fifty times, letter by letter, by which time the network was reproducing 95% of the sounds correctly. Many of the mistakes it made were not catastrophic because even though it often chose wrong sounds they resembled the correct ones and only deviated in, for example, voicing. The mistakes had a resemblance to those made by an English-speaking child learning to talk. It looks as though the things that cause problems for people also cause difficulties for networks.

The network's ability to generalize was tested by getting it to speak the next 439 words of the story, of which it managed 78% of the sounds correctly. The network appeared to have grasped the general principles of pronunciation in English rather than having learned each individual word as a separate case. When the output was passed through a speech generator, the words were clearly distinguished from each other, and it was also possible to understand what the child said in the last part of the text.

COMPUTO, ERGO SUM?

I compute, therefore I am! Is *being* the same as having a superb portable computer inside one's skull? Is the computer the banal underlying secret of mind and soul?

Theories about the "mechanism" of the brain have always been linked to the current status of technology. During the industrial revolution, ideas of the brain's function were based on cogwheels and pulleys, and with the coming of the telephone it was viewed as a gigantic switchboard. Since the 1950s people have thought of the brain as an enormous symbol-manipulating digital computer, and now it is beginning to be seen as a massively parallel analog computer.

Intelligence is a difficult phenomenon to come to grips with, if only because quite a lot of it is required even before the concept emerges.

Had we all wings like Raphael's angels, the Wright brothers' pioneering 50-meter flight at the soft sand dunes of Kitty Hawk wouldn't have been anything to write home about. However, their little hop, despite its primitive nature, was the first successful imitation of birds' free flight, and a boundary had been crossed once and for all. Our inability to fly makes us impressed, out of all proportion, with what is, after all, a very primitive imitation.

On the other hand, however, we tend to regard a small amount of intelligence as pure stupidity. Human intelligence exceeds any other known intelligence, and thus it is difficult to acknowledge a fruitfly's intelligence even though it is much more gifted than a shovel.

COMMUNICATION

Man's ability to communicate with the help of language frequently obscures the fact that the basic cognitive abilities are nonverbal. When one has driven unscathed through rush-hour traffic, one frequently wonders how it is possible to drive a car while thinking about such completely different things as how the day has gone, supper, the kids in school, one's bank overdraft, and sex. There is something meditative about driving, and it is incomprehensible that the numerous vital decisions which driving comprises do not impinge more on our consciousness than is actually the case.

Verbal, symbolized thinking is only the tip of the iceberg, the bulk of which is nonverbal cerebral activity. It was very late in evolution that the brain developed a strong verbal capacity. But since then, verbal communication has acted as a catalyst for social interaction between vast numbers of people.

Verbal communication is in fact not very rapid. One has to be an Italian sports commentator to produce more than ten words per second. This corresponds roughly to 100 bits per second. In comparison, the overwhelmingly nonsymbolic information we receive visually contains far more.

For example, the bandwidth of a television channel is approximately 1000 times greater than the bandwidth of the accompanying sound. The amount of information transmitted by the picture is 1000 times greater than that of the sound. The old saying that "one picture is worth a thousand words" is well-rooted in reality.

The division between the oral and the visual is, to all intents and purposes, the division between the symbolized and the nonsymbolized. The words of language are symbols of objects or concepts, but without physical relation to them, whereas pictures on a screen are physical reconstructions of the light that actually was emitted or reflected from an object at the time the pictures were taken. While we have subtitles translating foreign-language speech, there are no equivalent "sub-pictures" translating foreign images into a special national form. The whole idea of television is to reconstruct the light emissions recorded by the camera as faithfully as possible. There is no fundamental difference between "natural" and "artificial" light. When color television arrived, the reconstruction was better than before, and when it becomes holographic, it will be almost perfect.

Verbal communication is not particularly precise either. It can be very difficult to find the correct word to express the intuitive meaning one wishes to convey. Many discussions go to pieces because we cannot agree what a particular word means or how it should be used.

The brain is a slow symbol processor. When we think in sentences, the speed is not substantially greater than when we speak. It is not unusual for our vocal chords to move while we think, and sometimes we actually think aloud. The majority of our thinking activity does not follow verbal patterns, but occurs in brief images, which the brain has a hard time producing subtitles for when we speak.

Evolution has furnished us with a camera, two in fact, but no screen. If we had a light-emitting screen on our foreheads, communication would be considerably more precise than it is because we could present each other directly with pictures of our thoughts. Neurally controlled light emission certainly exists in nature, in fireflies for example, and in principle there is nothing in the way for constructing a biological light-emitting screen, even in color; it simply hasn't happened.

Hearing and speech did not arise simultaneously, and it could be the same with light-assisted communication. Just because we can't see the hands moving, it does not mean that evolution's clock has stopped, and it could be that the light screen is one of the delightful little surprises that Nature has in store for our remote descendants.

Actually, we do have a kind of screen in the head, namely our face with its around one hundred individual muscles. The facial expressions and the color changes associated with emotional experiences, for example embarrassment, allows us to transmit considerable amounts of information outside the narrow channel of speech.

INTUITION

A symbol has no direct connection with what it represents. The word "house", for example, has no connection with the buildings it describes. The arbitrary nature of the form of verbal symbols is also reflected in the widely differing symbols used for the same object in different languages.

How is the meaning of these random symbols arrived

at? What, in fact, is the meaning of *meaning*?

To start with, none of the symbols we are bombarded with has meaning for us. However, with time, our brain comes to extract meaning from the information it receives. The word "mother" gains meaning through the contact and interaction we have with her — through what she is, and what she can do. We knew who mother was before we knew the word. In the beginning "the Word" was, in fact, not!

Throughout life we learn to connect words with a certain number of concepts. Few people know more than 50 000 words and most manage their daily lives with just a few thousand. We understand many more things than we have words for. It is thus far from easy to give a detailed verbal description of, for example, the Pope's face. Even so, we recognize it intuitively and immediately from 5 billion others.

Recognition of a face is an example of symbol-free calculation. Many "natural" calculations are carried out without the help of symbols. The Moon doesn't work out its orbit in symbols, it just moves along it. When we bend an arm, we don't make a symbolic calculation either. We just do it.

All this nonsymbolic information processing is kept hidden from us. In the same way that the Moon doesn't "understand" that it is utilizing Newtonian motion, so are we unaware of how an associative look-up takes place in our memory. We don't experience the dynamics of the brain's neural network when we recognize a face. What we perceive is *the result* of the calculation process, not its course.

Intuitive thought is symbol-free and is much richer in information than symbolized reasoning. The wealth of information makes it difficult to transfer its products to narrow symbols which can be communicated. When a child learns to ride a bicycle, it is told to stop thinking about what it is doing and just do it. Intuition suffers from claustrophobia when hemmed in by words.

At first glance, it would seem futile to compare our thinking process with the Moon's orbital motion. The Moon is a deterministic system, the future state of which is determined entirely by the present state. Our cerebral processes, in contrast, give us the impression of free will.

It is immediately apparent when performing a look-up in our associative memory that free will is not operative. It is simply impossible to *stop* the recognition of the Pope's face once we have seen it. One can play stupid and *pretend* not to recognize a person, but that only happens after recognition has occurred.

Whether we have free will at all is another matter. We feel that we have, but it must be emphasized that there certainly are important mental processes over which we have no control whatsoever. Just as a policeman in court cannot refrain from letting the judge know that the person in the dock is "known to the police", so our associative memory cannot avoid telling us that a known face is recognized.

Neurons have themselves no free will and are not individually under the control of our free will. It is a great mystery how the psychological perception of free will takes place in a brain composed of neurons which are not free, but are, equally, not under control.

Maybe it is possible to view the problem from a computational angle. Because of the immense exchange of influence in the brain's neural network, no individual neuron or group of neurons can take complete control.

The neural decision process that makes us "free" is the collective effect of billions of neurons' mutual interaction. The multiplicity of interactions and the vast quantity of stored memories make the brain macroscopically unpredictable, even though all its microscopic neural processes are deterministic. The human mind is, like a universal computer, computationally irreducible, which means that the only way we can find out what a person will do in the future is to wait and see. There is no known shortcut.

The paradox of free will arises from the fact that the intuitive nonsymbolic processes which contribute to our decisions are inaccessible to our consciousness. We regard ourselves as "free" because in the end we are computationally irreducible, even when we try to look into ourselves. We are as little able to predict what precise decisions we shall take in future as anybody else. We can only wait and see.

CALCULATIONAL PARADIGMS

Calculations take place in macroscopic physical systems, whether we are talking about brains or computers. Computers come in two varieties — digital and analog — whereas brains at the moment at least come only in the analog model.

The characteristic of the digital computer is that its information processing is exclusively symbolic. At the lowest level which affects the information processing the machine only manipulates the binary symbols 0 and 1. No programmer is bothered about what happens below that level. In contrast, the builders of computers, electrical engineers, are principally concerned with the level below the digital, inasmuch as their main concern is to get a continuous physical system to behave digitally.

The analog computer uses physical processes directly to produce symbol-free calculations. For example, it is possible to construct an electric circuit whose current intensities will calculate the Moon's orbit in a given time frame. The physical quantities are usually continuous, which gives the analog computer a much broader spectrum of possible states than its digital cousin. Its calculations cannot be tabulated, not even in principle, like those of the digital computer which can only perform a finite number of calculations. The analog computer, however, can be constructed with much fewer hardware components. With a few transistors one can in the analog fashion perform complex mathematical operations such as multiplication and evaluation of logarithms. The same operations in a digital computer require hundreds of transistors.

On the other hand the analog computer is rarely particularly flexible. It has to be rebuilt to each new calculational task. The digital computer derives its flexibility from a separation of the physical and computational levels. The two binary symbols 0 and 1 mark the borderline between the two levels. Binary symbols are not restricted to a particular use, but can freely represent every binary pair of concepts. The digital mode has the advantage of decisively improving the logical stability of calculations, but this is at the expense of processing speed and informational content.

Digital computers cannot exist without an underlying analog level, but are always superstructures to real physical systems. Analog computers exist directly as physical systems and require no level separation. They are, however, difficult to program.

The neural network is a kind of analog computer which does not have to be rebuilt for each new set of calculations. It is not "programmable" in the usual sense, but can be influenced — trained — to perform the desired computation. Our brains contain neural networks, not digital computers, because Nature did not find it necessary to go the long way round and internally symbolize the basic cognitive functions.

In the same way that the digital computer is a symbol-processing superstructure for the analog computer underneath, man — possibly the only living organism to do so — has developed a symbol-processing superstructure on top of the brain's analog neural network. Not only has man's use of symbols turned him into a frightful chatterbox, but it has also made him believe falsely that the symbol-processing part of the brain is responsible for virtually all human information processing.

The symbol-processing superstructure is, naturally, part of the brain's neural network, but is, in the same way as the digital computer, forced into the straitjacket of symbols. Because evolution seldom scraps anything which works, the superstructure is a supplement to the unsymbolized, intuitive thought processes instead of being, as in the case of the digital computer, a limitation to it. Over several million years of evolution, these two aspects of the brain's function have been fused into a formidable whole that combines the best of two information processing paradigms.

SIMULATION

Simulating is not the same as *being*. A simulating person is, in fact, someone who swindles with his personal skills, and one usually runs into such people in front of draft boards, in the loan departments of banks, and at all-night poker games. We make sharp distinctions between real doctors and quacks, real aristocracy and catchpenny counts, because we fear that the simulation isn't perfect. Once in the United States a president simulated honesty to the extent that he had to go on national television to state: "I am not a crook" when he was about to be found out. In Austria it has been very difficult to decide whether a ruling president and former general secretary of the United Nations is simulating a clean conscience or whether he in fact did not participate in mass murder. In other countries, the cardboard counts and countesses tend to wind up in jail for stealing the cash box, rather than for pretending to be what they are not. The perfect simulator is someone who can carry off the deception without ever being suspected, and, for obvious reasons, we don't know how many of them there are.

While the analog computer actually *exists*, the digital computer is only a simulation, because it has an analog substructure. There is not a single digital computer anywhere in the world — only swindlers with false papers. The simulation of the digital computer is so perfect that it is very seldom unmasked. Swindlers are usually only discovered when they make mistakes which bring to light the truth about their internal constitution. Similarly calculation errors occur in digital computers when the physical degrees of freedom run away with them, such as when the air-conditioning system is not working properly.

One can also simulate an analog computer with a digital computer. With currently available components it is much cheaper to build a reliable semblance of an analog computer in this way than to build it from scratch. Naturally, this shortcut has its drawbacks in the form of longer processing time, but the advantages of precision and cost often outweigh the loss of speed.

Thus, layer upon layer of simulation can be produced. If the simulation is perfect, then the different layers are completely separated from each other. They correspond, more or less exactly, to the often nearly disconnected conceptual levels to be found in the scientific world. Physicists now speculate whether there is a more fundamental digital level beneath the apparently continuous levels of matter. If this were the case the basic structure of the universe that we live in would be that of a giant digital computer.

The renaissance which has taken place in recent years in the sphere of artificial neural networks had almost exclusively been based on digital simulation. These digital simulations are very time consuming and make enormous demands on computer power, but one may confidently expect these drawbacks to be entirely eradicated when, in the near future, electronic neural networks become generally accessible.

SYMBIOSIS

In man, the fusion of the digital and analog paradigms

has already occurred. Man has been outstandingly successful because he is the first living creature to combine intuitive, nonsymbolic computing with reasoned, symbolized calculation.

The serial nature of symbolized reasoning is directly related to the serial nature of our organs of speech. Our ears can analyze time variations in sound signals, while our sight analyzes both the spatial and temporal variations in the light we receive. Furthermore, both eyes and ears can compare two signals from the same source, giving us directionally sensitive hearing and depth-perceptive vision.

The speed with which we are able to process symbols is of the same order of magnitude as that with which our vocal organs operate, whereas the speed at which our intuition takes flight is much more like that of our image-processing ability. When thoughts take off they are not hampered by some ponderous "text machine" producing "subtitles" for the countless images flashing through our minds.

Reasoning is normally progressive. Logical and mathematical proofs are the best examples of this stepwise progression. Information is discarded along the way and the process is not immediately reversible. Furnished with an answer, one cannot trace back to the question. The human situation is often the opposite, we need to find causes for the events we experience. Most often we must consider a vast number of possibilities and a serial, reasoned analysis of causes would be altogether too slow. This is where our intuition steps in and allows us to review simultaneously an enormous number of conditions that limit the possible causes. It is this simultaneous consideration of the many factors which points us in the right direction for a solution — not serial consideration of them one at a time.

Without intuition our thinking would be as barren as the functioning of a digital computer. However, without symbolized reasoning we would be unable to communicate our thoughts in detail.

Similarly, the two calculational paradigms will not exclude each other in the electronic version. The first electronic, neural networks now on the market can be built into a digital computer so that it handles the network's input and output. There is thus a form of symbiosis between the two calculational modes which appears to be tailor-made for the demands that symbolic and nonsymbolic information processing impose.

Thus, the artificial neural network will provide the computer with a certain measure of intuition. This intuition, like that in the human, will be characterized by the capacity for generalization, nonsymbolic rule-formation and associative pattern recognition. The massive, inbuilt parallelism will permit the processing of enormous quantities of data in a very short time. The artificial neural network will not render the rigorous symbol-processing serial computer obsolete, in the same way as the human symbol-processing consciousness has not eliminated the soft symbol-free intuition.

In the symbiosis of these two aspects of computation lies the hope for the understanding of natural as well as artificial intelligence.

AFTERWORD

A COMPUTER'S DESTINY

Willem Klein was born in Amsterdam in 1912. His father was a doctor and ruled the Jewish family with an iron hand. Wim was somewhat gnomelike in appearance, with a large head and a small body. His brother Leo was the favorite of both the father and the mother. Wim was not very popular with the teachers at school either, but in high school he found a math master who recognized his special gifts.

When Wim was seventeen his mother committed suicide by throwing herself from an attic window, and was lying in the street when Wim came home from school. After that, the house was run by a woman who showed Wim much more affection than his mother had. This woman and her husband were quite unpretentious people and became almost foster parents to Wim and Leo, particularly after the death of their snobbish father in 1937.

Both Wim and Leo had been good with numbers from a very early age. They trained each other constantly by multiplying and dividing all the numbers they encountered; square roots and prime numbers were particular favorites. Wim was the sharper of the two and demonstrated his skills publicly for the first time on Dutch Radio in 1936. In 1932 his father had forced him to study medicine and held him to it until the father's death in 1937. The brothers shared a small inheritance and Wim began a serious attempt to drink his share away. He had already started drinking secretly while still young, but his father had a damping effect on the excesses. It was during this period that his studies literally dissolved.

Shortly after the Occupation and the beginning of the persecution of Dutch Jews, the Germans confiscated the remains of his legacy. In 1941, Wim enrolled at an Amsterdam hospital with the intention of completing his medical studies in three years. He never became a doctor. In 1943 Jews were forbidden to get a higher education and he was employed as a medical orderly at the Jewish Hospital. This afforded him slight protection from persecution, but his brother was deported and later died in Bergen-Belsen.

Wim managed to avoid arrest only by faking suicide, and his admission to a psychiatric clinic again gave him a brief period of safety. The beginning of 1944 found him working at the Jewish Hospital once more but, by the end of the year, he had to go underground. He remained in hiding in Amsterdam until the end of the war and was one of approximately 5000 Jews who survived in a city with a pre-war Jewish population of 120 000 plus.

Immediately after the war he began his career as a café performer. Equipped by his agent with a turban, a false moustache and the name Ali Ben Achmed, he performed feats of artistic computing in cheap eating houses. As early as the end of 1945 he had changed his agent and been rechristened Pascal after the French mathematician and began to work better venues as a member of a then-popular variety group. By the middle of 1946 he was with an act called "The Three Wise Men", made up of a mindreader Indra, a magician Iberko, and the numerical genius Pascal. For the next few years they worked the European circuits, mostly in Belgium.

By 1948, he was in Paris working the streets. For a couple of years he existed as an alcoholic bum, broken only by a brief spell as a pimp in Pigalle, after which he was deported from France as an undesirable alien.

After a love affair, his only relationship with a woman, in the Belgian town of Mons, he returned to Amsterdam at the beginning of the 1950s.

Here, his step-family noticed a small advertisement from the Center for Mathematics wanting people with mathematical ability for numeric calculations. He was quickly engaged and spent the next few years engaged in a wide diversity of jobs. He quickly bored of this life and in 1952 returned to France, where he met with great success traveling the schools of the country as a visitor, teaching people how to calculate. He was invited to England by the BBC to appear on television, and there he remained working as, among other things, a calculator demonstrator, appearing frequently on TV until 1955. Back in Amsterdam, he again worked at the Mathematics Center, but still found it monotonous.

In 1956, he met Professor Bakker, then director of the infant nuclear research center CERN in Geneva. In 1958 he was employed in the theoretical department to help physicists with their calculations. The excellent salary he received allowed him to pursue his two passions: calculations and alcohol. Until the mid-60s he played an important role as a computer, but saw himself being outdistanced by the new machines being developed in the USA and which were rapidly acquired by CERN. Over the next few years he tried to become a programmer, but his heart was not in it. As he said, computers have neither intelligence, intuition or knowledge and are not much fun as drinking buddies.

At the start of the 1970s he entertained the CERN summer students and it was quickly apparent that everyone working there loved his performances. He also began his phenomenal public "root extractions" which put him in the *Guinness Book of Records*, an achievement which gave him enormous pleasure. He continued his displays of prowess until his retirement in 1976.

Back in Amsterdam as a pensioner, he led an active life. The human computer Wim Klein was brutally stabbed to death in his own home on August 1, 1986. The murder has never been solved.

FURTHER READING

Artificial neural networks have by now a comprehensive body of literature. The following books, journals and articles have been of particular use to us.

H. KUHN
Wim Klein — Genie, Clown oder Wissenschaftler
Ted Siera Verlag, Hamburg, 1983

A fake autobiography of a real person written by a staff member at the European nuclear research organization CERN. The book is a personal and lively account of the life of an extraordinary human being.

DÆDALUS
Artificial Intelligence
Journal of the American Academy of Arts and Sciences
Winter 1988, Vol. 117, no. 1

This collection of articles is penetrating and interesting. The contributions are written by the leading scientists in the field of artificial intelligence. Several of the articles also treat the philosophical implications of neural networks.

D. E. RUMELHART, J. L. MCCLELLAND AND THE PDP-GROUP
Parallel Distributed Processing, Vols. 1, 2
The MIT Press, 1986

These best-selling books contain the basic theory and applications of neural networks of many different architectures and have contributed greatly to the dissipation of knowledge about neural networks. The books consist of articles written by members of the PDP-group and are edited by the two main authors who are psychologists. The books give an easy access to many different models of information processing with neural networks in a variety of cognitive problem areas and are indispensable to beginners in the field.

J. L. MCCLELLAND AND D. E. RUMELHART
Explorations in Parallel Distributed Processing
The MIT Press, 1988

A more recent handbook with models, programs and exercises in the techniques behind neural networks. The book also contains two diskettes with the source code for the exercises written in the programming language C.

T. KOHONEN
Self-Organization and Associative Memory
Springer-Verlag, 1988

T. KOHONEN
Content-Addressable Memory
Springer-Verlag, 1987

Teuvo Kohonen from Finland is one of the leading figures in the research on artificial neural networks and has written several books containing mathematical formulations of many different models. In particular he has worked with aspects of associative memory and the so-called self-organizing feature maps.

J. SEARLE
Minds, Brains and Science
Harvard University Press, 1984

A provocative book about the philosophical problems in neuroscience, but contains nothing specific about artificial neural networks.

M. L. MINSKY AND S. A. PAPERT
Perceptrons
The MIT Press, 1988

A classic that has had great importance for the research on perceptrons in both negative and positive directions. The new edition is a reprint of the edition from 1969 with handwritten comments. It contains an epilog written in 1988, defending the negative attitude towards the subject which this book originally caused.

B. A. HUBERMAN (ED.)
The Ecology of Computation in
Studies in Computer Science and Artificial Intelligence
Vol. 2, North Holland, 1988

A collection of articles written by persons with a broad background in artificial intelligence. The articles deal with problems in open systems where many different independent decision makers interact with each other. This book is the first to point out the very exciting possibilities for understanding new collective effects that may arise in large scale networks of real complex computers and not just in neural networks where the computing units are extremely simple.

W. DANIEL HILLIS
The Connection Machine
The MIT Press, 1985

This is the author's Ph.D. thesis and has become a classic. It describes the architecture of the first Connection Machine, which Hillis himself designed, and also tells how this machine is programmed. The book is relatively technical and can be hard to understand if one is not familiar with computer jargon.

R. M. J. COTTERILL
The Cambridge Guide to the Material World
Cambridge University Press, 1985, 1989

A thorough and easily comprehensible review of the

properties of matter from the elements over proteins to living tissue. Without the use of mathematics it describes many of the advances made in physics, chemistry and biology concerning the understanding of Nature at the microscopic and atomic levels. The new edition has been extended with material about neural networks.

E. R. KANDEL AND J. H. SCHWARTZ
Principles of Neural Science, Second Edition
Elsevier, 1985

A comprehensive text for medical students about the anatomy and physiology of the brain with emphasis on signal processing, membrane properties, sensation and perception.

C. H. BENNETT AND R. LANDAUER
The Fundamental Physical Limits on Computation
Scientific American, p. 38–46, July 1985

R. LANDAUER
Dissipation and Noise Immunity in Computation
Nature, Vol. 335, 779–784 (1988)

A. HODGES
Alan Turing: The Enigma of Intelligence
Counterpoint, 1985

An exciting book about the mathematician and computer

theoretician Alan Turing's life (1912-54). The book reviews Turing's work on the abstract mathematical theory of serial information processing. This among other things resulted in the description of a hypothetical universal machine which in principle can perform all digital computations. Turing mainly did this work before the Second World War when he and many other colleagues worked hard to break the German military codes. He claimed very early that the basis of intelligence is nothing but computation and that computers with time would be able to imitate cognitive behavior which would be very hard to distinguish from similar human behavior.

F. CRICK AND G. MITCHISON
The Function of Dream Sleep
Nature, Vol. 304, p. 111–114 (1983)

F. CRICK AND G. MITCHISON
REM Sleep and Neural Nets
The Journal of Mind and Behavior, Vol. 7, p. 229–249 (1986)

W. S. MCCULLOCH AND W. PITTS
A Logical Calculus of the Ideas Immanent in Nervous Activity
Bulletin of Mathematical Biophysics, Vol. 5, p. 115–133 (1943)

J. J. HOPFIELD
Neural Networks and Physical Systems with Emergent Collective Computational Abilities
Proceedings of the National Academy of Sciences, USA
Vol. 79, p. 2554–2558 (1982)

J. J. HOPFIELD, D. I. FEINSTEIN, AND R. G. PALMER
"Unlearning" has a stabilizing effect in collective memories
Nature, Vol. 304, p. 158–159 (1983)

T. J. SEJNOWSKI AND C. R. ROSENBERG
NETtalk: A Parallel Network that Learns to Read Aloud
Complex Systems, Vol. 1, 145–168 (1987)

W. D. HILLIS
The Connection Machine
Scientific American, Vol. 256, 86–93 (1987)

Neural Networks
Pergamon Press, Issued 1988 (quarterly)

This journal is the official organ for
The International Neural Network Society.

Neural Computation
The MIT Press, Issued 1989 (quarterly)

International Journal of Neural Systems
World Scientific Publishing Co, Issued 1989 (quarterly)

We have here mentioned only a few from the flood of articles that have followed in the wake of the new wave of interest in neural networks. One of the problems in following the development in this cross-disciplinary field is that the publications are scattered over a large number of different journals. The above-mentioned articles concern some of the examples described in this book.

INDEX